"人类非物质文化遗产代表作——二十四节气"科普丛书

中国农业博物馆 组编

Popular Science Book Series on
A Masterpiece of the Intangible Cultural Heritage of Humanity — The Twenty-Four Solar Terms

The Twenty-Four Solar Terms

Compiled and Edited by China Agricultural Museum

中国农业出版社
China Agriculture Press

北京 Beijing

丛书编委会
Editorial Board of the Book Series

主任 陈 通　刘新录
Directors

副主任 刘北桦　苑 荣
Deputy Directors

委员（按姓氏笔画排列 Ranking by Surname Strokes）
Members
王应德　王晓杰　王彩虹　毛建国　玉 云　皮贵怀
吴玉珍　吴德武　陈 宁　陈红琳　周晓庆　俞茂昊
唐志强　谢小军　巴莫曲布嫫

主编 苑 荣
Editor-in-Chief

副主编 唐志强　韵晓雁　王晓鸣　程晋仓
Deputy Editors-in-Chief

翻译 徐立新
Translator

本书编委会
Editorial Board of This Volume

主编 王应德　周晓庆
Editors-in-Chief

副主编 王晓鸣
Deputy Editor-in-Chief

参编人员 张 苏　陶妍洁　李毅强　付 娟
Editorial Staff

摄影 杨 晋　张 彧　陈跃敏　陈广程　郭振毅
Photographers
张争鸣　刘清晖　彭徐蒙　唐寒飞　耿洪杰
金代江　贺敬华　韩翠芝　李冬梅　尹 涛
周 峻

绘画 林帝浣
Illustrator

"人类非物质文化遗产代表作——二十四节气"科普丛书
Popular Science Book Series on
A Masterpiece of the Intangible Cultural Heritage of Humanity — The Twenty-Four Solar Terms

序
Foreword

中华文化，博大精深，灿若星河，传承有序，绵延不绝。作为人类非物质文化遗产代表、凝结中华文明智慧的"二十四节气"在我国自创立以来，已经传承发展2 000多年。它是中国人观天察地、认知自然所创造发明出的时间知识体系，也是安排农业生产、协调农事活动的基本遵循，更是中国社会顺天应时、指导实践的生活制度。它是中华优秀传统文化中文明成果的典型代表，体现了传统农耕文明的智慧性，彰显了中国人认知宇宙和自然的独特性及其实践活动的丰富性，凸现了中国人与自然和谐相处的哲学思想、文化精神和智慧创造。

Chinese culture, wide-embracing and profound, brilliant in numerous fields, inherited in an orderly manner, has been developing without interruption. The "Twenty-Four Solar Terms" is a masterpiece of the intangible heritage of humanity and a crystallization of Chinese civilization and wisdom for 2,000 years of its existence. It is a time knowledge system invented by the Chinese people to observe heaven and earth and to learn about nature, the basic principles to organize agricultural production and coordinate farming activities, and also the life system for the Chinese society in its conformation to natural and meteorological timings and in its guidance of daily practices. As the typical representation of the fruit of the best of the traditional Chinese civilization, it embodies the wisdom of traditional farming civilization, reflects the Chinese people's unique interpretation of the universe and nature and the rich practices therein, and highlights their philosophical concepts, cultural spirit and intelligent creativity in their harmonious with nature.

"二十四节气"起源于战国时期,在公元前140年就已经有完整的"二十四节气"记载。从时间上,作为太阳历,早于儒略历(公元前45年)近一个世纪。"二十四节气"较之公历更准确地标识了地球视角的太阳运行规律。农谚就"二十四节气"同公历的关系说道:"上半年是六廿一,下半年来八廿三。每月两节日期定,最多不差一两天。"这里所说的"不差",不是"二十四节气"不准,而是公历有"差"。我们的生活要顺天应时,生活在自然体系之中,就应该把自己看成是包括自然界在内的客观世界的组成部分。无限制地扩大人的能力,破坏自然规律,其后果是难以意料的。

"人类非物质文化遗产代表作-----二十四节气"科普丛书
Popular Science Book Series on
A Masterpiece of the Intangible Cultural Heritage of Humanity — The Twenty-Four Solar Terms

The "Twenty-Four Solar Terms" originated in the Warring States Period. There were already complete records of the "Twenty-Four Solar Terms" in 140 B.C., almost one century earlier than the Gregorian calendar (45 B.C.), also a solar calendar. It is more accurate than the latter in indicating the laws of the sun's movement from the earth's perspective. Agricultural proverbs identify its relationship with the Gregorian calendar: "In the first half of the year, the solar terms fall on the 6th and 21st of each month, and in the second half, they fall on the 8th and 23rd. There are two solar terms in each month, with an adjustment of one or two days." Here, "adjustment" is not the result of inaccuracy of the "Twenty-Four Solar Terms" but errors on the part of the Gregorian calendar. As we need to conform to the natural and meteorological laws and live in natural systems, we should regard ourselves as components of the objective world, including the nature. Expanding human capacity without constraint and disrupting the natural laws may lead to unexpected consequences.

对中国人来说，"二十四节气"是我们时间制度整体的一部分，它是指导我们包括农业在内的创造生活资料的一切活动的时间节律。而我们的情感表达、礼仪等调节人际关系、社会关系的活动则以对月亮运行周期观察为基础的太阴历为节律。我们的传统节日体系大都是以太阴历为依据的。在我们的阴阳合历的整体框架里认识"二十四节气"，领会我们的先辈以置闰的办法精妙恰当地协调二者的对应关系，体现了中华传统文化的精奥和人文精神。

For the Chinese, the "Twenty-Four Solar Terms" is part of our time regime, the time prosody that directs all activities that produce living materials, including agriculture. The time prosody of activities that accommodate interpersonal and social relations, such as our emotional expression and etiquette and protocol, is the lunar calendar based on observations of the moon's periodic movement. Our system of traditional festivals is mostly founded on the lunar calendar. It is advisable to interpret the "Twenty-Four Solar Terms" in the overall framework of the lunisolar calendar, and understand how our ancestors wisely and aptly coordinate the correspondence between the solar and lunar calendars by means of intercalation, which reflects the beauty and humanistic spirit of traditional Chinese culture.

"二十四节气"融合四季,贯穿全年,广为实践,流布全国,影响世界。其作为我国优秀传统文化的典型代表和人类非物质文化遗产代表作项目,富含中国人特有的哲学思想、思维理念和人文精神,具有广泛的参与度和社会影响力,引发世人的关注与探索。2016年11月30日,在文化部非物质文化遗产司指导下,在中国民俗学会支持下,由中国农业博物馆作为牵头单位,联合相关社区单位申报的"二十四节气——中国人通过观察太阳周年运动而形成的时间知识体系及其实践",被联合国教科文组织列入人类非物质文化遗产代表作名录。这是中国非遗保护工作取得的一项重要成果,也是对外文化交流的一次成功实践。在其带动影响下,全国人民乃至世界人民对"二十四节气"的认知、认同、参与和实践空前提高,进一步彰显和增强了中国人的文化自觉和文化自信。

The "Twenty-Four Solar Terms", integrating the four seasons and covering the whole year, is widely practiced throughout China, with influences on the whole world. As China's best

"人类非物质文化遗产代表作-----二十四节气"科普丛书
Popular Science Book Series on
A Masterpiece of the Intangible Cultural Heritage of Humanity — The Twenty-Four Solar Terms

representative of traditional culture and a masterpiece of the intangible heritage of humanity, it is full of philosophical thoughts, thinking patterns and humanistic spirit unique to the Chinese, enjoying a wide participation and social influence, and commanding attention and exploration from around the world. On Nov. 30, 2016, under the guidance of the Intangible Cultural Heritage Department of the Ministry of Culture and with the support of the China Folklore Society, "the Twenty-Four Solar Terms, knowledge of time and practices developed in China through observation of the sun's annual motion" submitted by the China Agricultural Museum together with related community organizations, was entered onto the list of Masterpieces of the Intangible Cultural Heritage of Humanity by the UNESCO. This was a significant achievement in China's intangible heritage protection, and also a successful practice in cultural exchange. Due to this endeavor and its influence, the people of China and of the world have unprecedentedly heightened their knowledge, identification, participation and practice regarding the "Twenty-Four Solar Terms", which further reflects and enhances Chinese people's cultural awareness and self-confidence.

出版这套"人类非物质文化遗产代表作——二十四节气"科普丛书,有助于在更大更广的范围和层面普及传播节气的相关知识,进一步增强遗产实践社区和群众的自豪感与凝聚力,激发传承保护的自觉性和积极性,扩大关于传统时间知识体系的国际交流与对话,推动人类文明交流互鉴。

The publication of this book series on *A Masterpiece of the Intangible Cultural Heritage of Humanity—The Twenty-Four Solar Terms* shall be conducive to its spread and popularization on a larger scale and in a wider sphere, further enhance the sense of pride and solidarity on the part of the inheritance practice communities and masses, inspire their awareness and initiative in preservation and protection, expand international exchanges and dialogues on traditional time knowledge systems, and promote exchanges and mutual learning between human civilizations.

期望并相信这套丛书能够得到社会各界人士的喜爱。

We sincerely hope and cordially believe that this series will win the hearts of readers of various circles.

谨为序。

Please enjoy your reading of this volume.

刘魁立
Liu Kuili

2019 年 3 月
March 2019

二十四节气
The Twenty-Four Solar Terms

前 言
Preface

  二十四节气是中国人认知一年中时令、气候、物候等方面变化规律所形成的知识体系和社会实践，是中国传统历法体系及其相关实践活动的重要组成部分。在国际气象界，这一时间知识体系被誉为"中国的第五大发明"。

The Twenty-Four Solar Terms is a knowledge system and social practice that the Chinese people have developed from their observation and study of the rules in the changes of the seasons, climate, and phenology. It is an important component of the traditional Chinese calendar system and related practices. In the international meteorological circle, this system of time has been reputed to be "the fifth great invention of the Chinese".

2016年11月30日,联合国教科文组织正式通过决议,将中国申报的"二十四节气——中国人通过观察太阳周年运动而形成的时间知识体系及其实践"列入联合国教科文组织人类非物质文化遗产代表作名录。

On November 30, 2016, by an official resolution of the UNESCO, "The Twenty-Four Solar Terms—knowledge of time and practices the Chinese developed through observation of the Sun's annual motion" was proclaimed a Masterpiece of the Intangible Cultural Heritage of Humanity.

目录
Contents

序
Foreword

前言
Preface

二十四节气概述
An Overview of the Twenty-Four Solar Terms —— 1

二十四节气由来
Origin of the Twenty-Four Solar Terms —— 7

立春 Beginning of Spring	13
雨水 Rain Water	18
惊蛰 Insects Awaken	23
春分 Spring Equinox	28
清明 Freshgreen	33
谷雨 Grain Water	38
立夏 Beginning of Summer	43
小满 Lesser Fullness	48
芒种 Grain in Ear	53
夏至 Summer Solstice	58
小暑 Lesser Heat	63
大暑 Greater Heat	68
立秋 Beginning of Autumn	73
处暑 End of Heat	78
白露 White Dew	83
秋分 Autumn Equinox	88
寒露 Cold Dew	93
霜降 First Frost	98
立冬 Beginning of Winter	103
小雪 Light Snow	108
大雪 Heavy Snow	113
冬至 Winter Solstice	118
小寒 Lesser Cold	123
大寒 Greater Cold	128
结语 Concluding Remarks	133
附录 Appendix	135

二十四节气概述
An Overview of the Twenty-Four Solar Terms

　　二十四节气是中国古人在长期的农业生产实践中，依据太阳在黄道（以地球为中心，太阳在恒星中的周年视路径）上位置的变化，总结我国一定地区（以黄河中下游地区为代表）在一个回归年中的天文、季节、气候、物候、农事活动等方面变化规律和特征的一种指导农业生产和人们生活的历法。

In their long-term agricultural production practices, the ancient Chinese had been observing the Sun's changes on the ecliptic (the Sun's visual annual path among the stars, with the earth as the center), and summarizing the rules of changes and features of certain regions (typically, the regions of the central and lower reaches of Yellow River) of the country in terms of the astronomy, seasons, climate, phenology and farming activities within a tropical year. The Twenty-Four Solar Terms have been serving as a calendar system in guiding their agricultural production and everyday life.

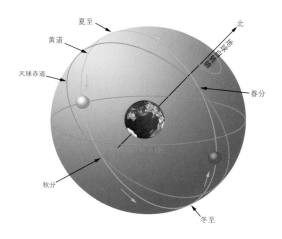

太阳黄道坐标系
The Solar Ecliptic Coordinates

二十四节气是将黄道假定为一个大圆，太阳从黄经 0°起，沿黄经每运行 15°所经历的时日称为"一个节气"。每年运行 360°，共经历 24 个节气，依次为：立春、雨水、惊蛰、春分、清明、谷雨、立夏、小满、芒种、夏至、小暑、大暑、立秋、处暑、白露、秋分、寒露、霜降、立冬、小雪、大雪、冬至、小寒、大寒（注：旧时是以冬至为一个完整节气体系的开始）。

The system of 24 Solar Terms works like this. The ecliptic is assumed to be a large circle. The sun starts from celestial longitude 0° and moves across the longitudes. The number of days covered for every 15° that the sun moves across is known as "one solar term". The sun moves 360° each year, covering altogether 24 solar terms. These 24 solar terms are, in their natural order of occurrence: Beginning of Spring; Rain Water; Insects Awaken; Spring Equinox; Freshgreen; Grain Rain; Beginning of Summer; Lesser Fullness; Grain in Ear; Summer Solstice; Lesser Heat; Greater Heat; Beginning of Autumn; End of Heat; White Dew; Autumn Equinox; Cold Dew; First Frost; Beginning of Winter; Light Snow; Heavy Snow; Winter Solstice; Lesser Cold, and Greater Cold. (Note: In old times, Winter Solstice was regarded as the start of a complete solar term system.)

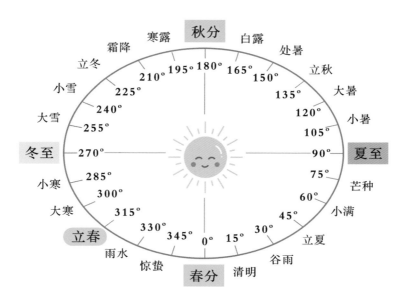

二十四节气黄道位置图
Positions of 24 Solar Terms on the Ecliptic

其中，反映四季变化的有立春、春分、立夏、夏至、立秋、秋分、立冬、冬至；反映温度变化的有小暑、大暑、处暑、小寒、大寒；反映天气现象的有雨水、谷雨、白露、寒露、霜降、小雪、大雪；反映物候现象的有惊蛰、清明、小满、芒种。每一个节气下，又细分为三候，每五日为一候，共计七十二候，反映了黄河流域地区一年中气候变化的一般情况。

Among them, the following reflect the changes of the seasons: Beginning of Spring, Spring Equinox, Beginning of Summer, Summer Solstice, Beginning of Autumn, Autumn Equinox, Beginning of Winter, and Winter Solstice.
The following indicate changes in temperatures: Lesser Heat, Greater Heat, End of Heat, Lesser Cold and Greater Cold.
These reflect meteorological phenomena: Rain Water, Grain Rain, White Dew, Cold Dew, First Frost, Light Snow, and Heavy Snow.
The following indicate phenological phenomena: Insects Awaken, Freshgreen, Lesser Fullness and Grain in Ear.
Each solar term is further divided into three pentads, each pentad composed of five days. There are altogether 72 pentads in a year. These display the general conditions of climatic changes throughout the year in the Yellow River basin.

由于二十四节气主要反映的是太阳的周年视运动，所以在公历中，它们的日期是相对固定的，上半年的节气在6日、21日，下半年的节气在8日、23日左右，前后不差1～2日。其中，每月第一个节气为"节气"，即：立春、惊蛰、清明等12个节气；每月的第二个节气为"中气"，即：雨水、春分、谷雨等12个节气。"节气"和"中气"交替出现，各历时约15天，现在人们已经把"节气"

和"中气"统称为"节气"。在传统历法制定中,中气是阴历设置闰月的重要依据,以没有中气的月份置闰,为上一月的闰月。

The twenty-four solar terms mainly reflect the annual visual movement of the sun. Therefore, in the solar calendar, the dates of the solar terms are relatively regular and fixed, on the 6th and 21st in the first half of the year, and on the 8th and 23rd in the second half of the year, or one or two days before or after. Among these, the first solar term in each month is called "*jieqi*" (initial point), i.e. the 12 solar terms such as Beginning of Spring, Insects Awaken, and Freshgreen. The second solar term of each month is called "*zhongqi*" (midpoint), i.e. the 12 solar terms such as Rain Water, Spring Equinox and Grain Rain. The initial points and the midpoints occur alternately, each lasting approximately 15 days. Nowadays people refer to "*jieqi*" and "*zhongqi*" collectively as "*jieqi*". In the evolution of the traditional calendar, "*zhongqi*" was an important basis for setting up the leap month, as the month without a "*zhongqi*" was the leap month of the previous month.

节气、中气一览表

月份(阴历月)	节气	中气
正月	立春	雨水
二月	惊蛰	春分
三月	清明	谷雨
四月	立夏	小满
五月	芒种	夏至
六月	小暑	大暑
七月	立秋	处暑
八月	白露	秋分
九月	寒露	霜降
十月	立冬	小雪
十一月	大雪	冬至
十二月	小寒	大寒

二十四节气
The Twenty-Four Solar Terms

An Overview of the Initial Points and Midpoints of Solar Terms in Each Month

Month (Lunar)	Initial Point (*jieqi*)	Midpoint (*zhongqi*)
First	Beginning of Spring	Rain Water
Second	Insects Awaken	Spring Equinox
Third	Freshgreen	Grain Water
Fourth	Beginning of Summer	Lesser Fullness
Fifth	Grain in Ear	Summer Solstice
Sixth	Lesser Heat	Greater Heat
Seventh	Beginning of Autumn	End of Heat
Eighth	White Dew	Autumn Equinox
Ninth	Cold Dew	First Frost
Tenth	Beginning of Winter	Light Snow
Eleventh	Heavy Snow	Winter Solstice
Twelfth	Lesser Cold	Greater Cold

二十四节气由来
Origin of the Twenty-Four Solar Terms

二十四节气的产生与地球的公转、自转有关，由于地球自转时，自转轴总是指向同一方向（北极星方向），所以地球自转轴和公转轨道平面始终保持着一个66°34′的倾角，这就使得地球在公转轨道不同位置时，同一地区接受光热的程度是不同的。

The origin of the twenty-four solar terms has to do with the rotation and revolution of the earth. When it rotates, its rotational axis is always pointed in one direction (North Star), and thus this axis always maintains an inclination angle of 66°34' with the orbit

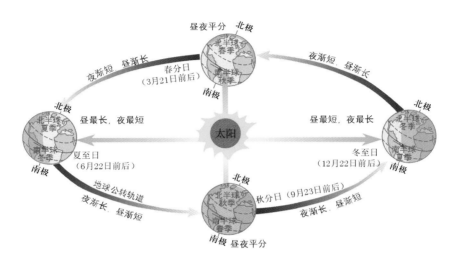

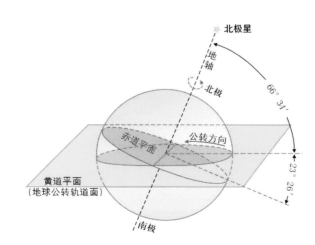

地球自转、公转示意图
Diagrams of the Earth's Rotation and Revolution

plane of the earth's revolution. This means that the earth receives a different degree of heat and light at different locations of the earth's orbit of revolution.

对这种地日关系所产生的规律性变化，中国古人在长期观察过程中认识不断深化，逐渐发明了以圭表测日影的方法，并在此基础上首先发现了冬至、夏至、春分和秋分，《尚书·尧典》中就记载有"日中""日永""宵中""日短"，分别对应"春分""夏至""秋分""冬至"。

As a result of long-term observation, ancient Chinese never ceased deepening their understanding of the regular changes in the earth-sun relationship, and gradually invented the method of measuring shadows with sun-dials, on whose basis they identified the winter and summer solstices, and the spring and autumn equinoxes. The "*rizhong*" (mid-day), "*riyong*" (day long), "*xiaozhong*" (mid-night), and "*riduan*" (day short) as documented in *Shang Shu · Yao Dian* (Book of Documents, Canon of Yao) correspond respectively to "Spring Equinox", "Summer Solstice", "Autumn Equinox", and "Winter Solstice".

《尚书·尧典》
Shang Shu · Yao Dian

春秋战国时期，《管子·轻重己》《左传·僖公五年》《吕氏春秋》等文献中记载的节气增加为8个，分别为立春、春分、立夏、夏至、立秋、秋分、立冬、冬至。这8个节气标示出季节的转换，清楚地划分出一年的四季。

In the Spring and Autumn Period and the Warring States Period, the number of the solar terms as documented in literature such as *Guan Zi · Qing Zhong Ji* (Writings of Master Guan, Economic Policies), *Zuo Zhuan · Xi Gong Wu Nian* (The Commentary of Zuo, Fifth Year of Duke Xi of the State of Lu), and *Lü Shi Chun Qiu* (Master Lü's Spring and Autumn Annals) has been increased to eight: Beginning of Spring, Spring Equinox, Beginning of Summer, Summer Solstice, Beginning of Autumn, Autumn Equinox, Beginning of Winter, and Winter Solstice. These eight terms indicate the change of seasons, clearly marking the four seasons of the year.

秦汉年间，二十四节气已完全确立。西汉时期《淮南子》一书第一次系统、完整、科学地记载了二十四节气的划分和观察计算方法，并且拥有与现代完全一致的二十四节气的名称，这也是中国历史上关于二十四节气的最早记录。

During the Qin and Han Dynasties, the twenty-four solar terms had been fully established. *Huai Nan Zi* (Writings of the Masters South of the Huai River) of the Western Han Dynasty, for the first time in history, systematically, comprehensively and scientifically documented the division and observation and computation of the twenty-four solar terms, and recorded their names in full consistence with what they are called in modern times. This was the earliest record of the twenty-four solar terms in Chinese history.

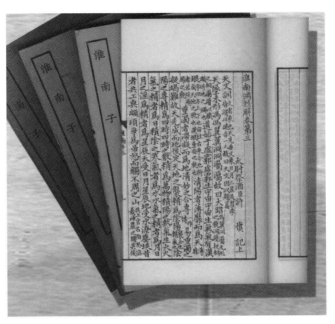

《淮南子》
Huai Nan Zi

汉武帝元封七年（公元前104年），由邓平、落下闳等制定的《太初历》，正式把二十四节气定于历法，明确了二十四节气的天文位置。

In the seventh year of the Yuanfeng Period in the reign of Emperor Wu of the Han Dynasty (104 B.C.), *Tai Chu Li* (Grand Inception Calendar) as formulated by Deng Ping, Luoxia Hong and others, officially included the twenty-four solar terms in the calendar, and identified their astronomical status.

至此,二十四节气作为中国人特有的时间制度,深刻影响和指导着人们的生产、生活,衍生发展了诸多有关天文、物候、农谚、音律、民谣、诗文词曲、饮食养生等的文明成果。春生、夏长、秋收、冬藏,围绕每一个时令节点,人们自发地组织农事生产,有序地安排日常生活,举办丰富多彩的节令仪式和民俗活动,形成了包括养生之道在内的独特的生活理念、风俗和方式。

Since then, as the distinct time system of the Chinese, the twenty-four solar terms have been fundamentally influencing and guiding Chinese people's life and work, and many fruits of civilization have derived, in terms of astronomy, phenology, farmers' proverbs, prosody, ballads, poems and lyrics, as well as diet and regimen. Sprouting in spring, growing in summer, harvesting in autumn, and hoarding in winter—around each solar term, people accordingly organize their production, arrange their daily life, and conduct joyous seasonal rites and folk events, thus developing their unique concepts, customs and ways of life, including their regimen principles.

立 春
Beginning of Spring

立春（彭徐蒙 摄）
Beginning of Spring (Photo by Peng Xumeng)

二十四节气
The Twenty-Four Solar Terms

　　立春，二十四节气中的第一个节气，公历每年 2 月 3 ~ 5 日，太阳到达黄经 315° 时。立春，意味着冬去春来，万象更新。

The Beginning of Spring, the first of the Twenty-Four Solar Terms, falls on February 3-5 on the Gregorian calendar, when the sun reaches celestial longitude 315°. The Beginning of Spring means the departure of winter and the advent of spring, when everything begins afresh.

　　古时将立春分为三候：一候东风解冻，二候蛰虫始振，三候鱼陟负冰。说的是东风送暖，大地开始解冻；五日后，蛰居的虫类慢慢在洞中苏醒；再过五日，河里的冰开始融化，鱼开始到水面上游动，此时水面上还有没完全融解的碎冰片，如同被鱼负着一般浮在水面。

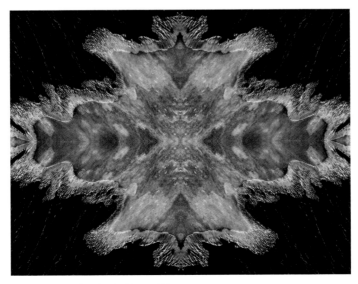

立春一候：东风解冻（杨晋 摄）
First Pentad of Beginning of Spring: East Wind Thawing the Earth (Photo by Yang Jin)

二十四节气
The Twenty-Four Solar Terms

In ancient times, the Beginning of Spring was divided into three pentads: in the first pentad, the east wind serves to thaw; in the second pentad, insects in hibernation begin to squirm; in the third pentad, fish start to surface with ice yet on their backs. This means that spring all starts from the warmth being brought along by the east wind and from the earth beginning to thaw. Five days later, insects are gradually waking up from their hibernation in their caves and holes. Five more days later, ice starts to melt in rivers and fish begin to swim here and there on the water surface. At this time, there are broken ice pieces that have not completely melted on the surface of water bodies, and it looks as if the ice slices are floating on the water surface on the backs of fish.

立春二候：蛰虫始振（杨晋 摄）
Second Pentad of Beginning of Spring: Insects Squirming (Photo by Yang Jin)

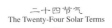

立春三候：鱼陟负冰（杨晋 摄）
Third Pentad of Beginning of Spring: Fish Swimming with Ice on Their Backs (Photo by Yang Jin)

时至立春，气温、日照、降雨趋于上升或增多，春的气息已经显露，提醒人们新的一年农业生产即将开始，春耕备播工作进入紧张阶段。

By the Beginning of Spring, the temperature, sunlight and rainfall tend to increase or rise; the breath of spring begins to appear, reminding people that the agricultural production for a new year will soon begin; and a busy phase of spring tilling and plowing and seed sowing preparations is launched.

一年之计在于春，立春在古代不仅是一个节气，还是一个重要的节日。官方要举行盛大的迎春祭祀活动，通过鞭春、报春等活动，预示春耕开始，以示丰兆，策励农耕。民间则有咬春、演春等习俗。

The whole year's work success depends on a good start in spring. The Beginning of Spring was not only a solar term, but also an important festival in ancient times. The government would sponsor grand ceremonies for sacrificial offerings and for ushering in the spring. Through such events as whipping the spring (bull) and announcing the spring, the timing of the start of spring plowing was conveyed to the whole society, the purpose being to symbolize a bumper harvest in a year of favorable winds and rains, and to encourage and give incentive to farming. Traditionally, there were such folk customs as biting the spring (turnips) and performing the spring (vendors of all trades carrying a colorful mansion miniature all the way to the office of provincial administration commissioners in fanfare and loud music to celebrate the Spring Festival one day beforehand).

立春时节，乍暖还寒，人体肌肤腠理变得疏松，人体内的正气抵御外部袭击的能力变弱，风邪易乘虚而入，容易导致风寒外感、风湿痹痛、头痛发热、咳嗽气喘等症状。此时，应注意养阳，提高身体免疫力。

At the Beginning of Spring, it may be warmer but then suddenly colder. The human skin and muscles become relaxed and loose, and the ability of the human body to defend itself against any external attacks becomes weak. The wind-evil may seize the opportunity to attack and slip in, so people suffer from coughs due to wind-cold pathogen, rheumatism and numbness, headaches and fever, coughing and shortness of breath. At this time, it is advisable to preserve and cultivate the *yang* energy, and enhance the body's immunity.

雨 水
Rain Water

雨水（陈跃敏 摄）
Rain Water (Photo by Chen Yuemin)

二十四节气
The Twenty-Four Solar Terms

　　雨水，二十四节气中的第二个节气，公历每年 2 月 18 ~ 20 日，太阳到达黄经 330° 时。此时，气温回升、冰雪融化、降水增多，故取名为雨水。

Rain Water, the second of the Twenty-Four Solar Terms, falls on February 18-20 on the Gregorian calendar, when the sun reaches celestial longitude 330°. At this time, the temperature rises, ice and snow begin to melt, and rainfall increases, which is why the solar term is so named.

　　雨水三候：一候獭祭鱼，二候鸿雁北，三候草木萌动。就是说这时候水獭开始捕鱼了，还要做出"先祭后食"的样子；五天后，大雁开始从南方飞回北方；再过五天，就能看到春雨霏霏、草木吐芽的早春气息了。

雨水一候：獭祭鱼（杨晋　摄）
An Otter Praying to a Fish Caught in the First Pentad of Rain Water (Photo by Yang Jin)

二十四节气
The Twenty-Four Solar Terms

The three pentads of Rain Water are like this. In the first pentad, otters pay homage to fish; in the second pentad, swan geese migrate to the north; and in the third pentad, plants start to bud. In other words, otters start to fish but display their catch on water margins instead of eating them up immediately, as if these are sacrificial offerings. Five days later, swan geese return from the south to the north. Another five days later, one can see the breath of early spring from the misty spring rain falling and from the first sprouting of grass and trees.

雨水二候：鸿雁北（杨晋 摄）
Swan Geese Migrating to the North in the Second Pentad of Rain Water (Photo by Yang Jin)

雨水三候：草木萌动（杨晋 摄）
Plants Starting to Bud in the Third Pentad of Rain Water (Photo by Yang Jin)

雨水之后，气温回升较快，雨量渐渐增多，各地树木、花草开始萌动，越冬作物开始恢复生长，一年一度的春耕生产即将开始。

After Rain Water, the temperature rises quickly, the rainfall gradually increases, trees, flowers and grass start to germinate, the overwintering crops resume growth, and the annual plowing and cultivation in the spring season is about to begin.

雨水节气川西地区有回娘家、拉保保、撞拜寄等习俗，取"雨露滋养易于生长"之意。客家地区则有通过爆炒糯谷米花，来占卜当年稻获丰歉的占稻色习俗。

二十四节气
The Twenty-Four Solar Terms

During the solar term of Rain Water, there are such customs in western Sichuan Province as wives visiting their own parents, and young parents asking the first encountered stranger as the sworn father or mother of their child. These customs symbolize morning dews nourishing one's growth. In the Hakka areas, there is the custom to foretell the year's rice harvest by examining the color of the stir-fried flakes of sticky rice.

雨水节气前后，正处于阴退阳长的转折期，此时人体经过一冬的收缩，开始变得舒展，毛孔也由封闭状态开始张开，因此在日常保健中，要注意防寒祛湿，养好脾胃。

Around the solar term of Rain Water is a transitional period in which the *yin* is declining and the *yang* is increasing. At this time, after a whole winter's contraction, the human body starts to relax and unfold, and the pores start to open up from the previous state of closure. Consequently, in daily health building, one needs to guard against cold and wetness, and to protect the spleen and the stomach.

惊蛰
Insects Awaken

惊蛰（杨晋 摄）
Insects Awaken (Photo by Yang Jin)

惊蛰，古称"启蛰"，二十四节气中的第三个节气，公历每年3月5～6日，太阳到达黄经345°时。惊蛰的意思是天气回暖，春雷始鸣，惊醒蛰伏于地下冬眠的昆虫。

Insects Awaken, also known as "Start of Insects" in ancient times, the third of the Twenty-Four Solar Terms, falls on March 5 or 6 on the Gregorian Calendar, when the sun reaches celestial longitude 345°. "Awakening of Insects" means that the temperature rises, and spring thunders start to sound, thus waking up the insects lying in hibernation under the ground.

惊蛰三候：一候桃始华，二候仓庚鸣，三候鹰化为鸠。意思是说桃花自此渐盛；五日后黄鹂开始鸣叫；又五日，鹰开始悄悄地躲起来繁育后代，而原本蛰伏的鸠开始鸣叫求偶。

惊蛰一候：桃始华（杨晋 摄）
Peach Blossoms Starting to Flourish in the First Pentad of Insects Awaken (Photo by Yang Jin)

It also has three pentads. In the first pentad, peach trees start to blossom; in the second, black-naped orioles begin to chirp; and in the third, hawks transform into turtle-doves. In other words, peach blossoms start to flourish first; five days later, one first hears the chirping of orioles; another five days later, hawks disappear because they hide themselves to hatch their eggs, and the turtle-doves that have previously been quiet are now beginning to chirp in pursuit of a love mate.

惊蛰二候：仓庚鸣（杨晋 摄）
Black-Naped Orioles Starting to Chirp in the Second Pentad of Insects Awaken (Photo by Yang Jin)

惊蛰三候：鹰化为鸠（杨晋 摄）
Hawks Transforming to Turtle-Doves in the Third Pentad of Insects Awaken (Photo by Yang Jin)

　　惊蛰在农业生产上是一个重要的节气，中国自古就把它视为春耕开始的日子，天子往往也在这个时候发布劝农耕种的诏书。

Insects Awaken is an important solar term in agricultural production. It has been regarded as the start of spring plowing. It was usually at this time that the Son of Heaven (the emperor) issued decrees to urge farmers to start plowing and cultivation.

二十四节气
The Twenty-Four Solar Terms

惊蛰节气，古时为了呼应天神的惊雷，有蒙鼓皮的习俗。广东等地有祭白虎、打小人的习俗，寻求心灵慰藉，祈求一年平安顺利。山西等地则有吃梨、除虫等习俗。

At this solar term, the ancient Chinese had the custom of smoothing out and sewing drumheads to coordinate with the alarming thunders from divine gods. In Guangdong Province and some other places, there were the customs of providing sacrificial offerings to painted tigers on paper, and beating the figure of the most hated person drawn on paper, in order to seek emotional comfort and pray for the year's success and auspiciousness. In places like Shanxi Province, there were customs of eating pears and getting rid of insects.

惊蛰时，人体的肝阳之气渐升，阴血相对不足，养生应顺乎阳气的升发、万物始生的特点，使自身的精神、情志、气血也如春日一样舒展畅达，生机盎然。

At Insects Awaken, the *yang* energy in the liver of the human body starts to rise, and the *yin*-blood is relatively insufficient. For health building, one should conform to the general rise of the *yang* energy and the overall trend of everything starting to grow, and relax and lift up one's spirit, sentiment, mood and *qi*-blood, all of which, like the spring sun, is smooth, all-embracing, and full of vitality.

春 分
Spring Equinox

春分（陈广程 摄）
Spring Equinox (Photo by Chen Guangcheng)

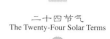

春分，二十四节气中的第四个节气，公历每年 3 月 21 日前后，太阳到达黄经 0°时。此时阳光直射赤道，南、北半球昼夜时间相等。春分"昼夜均而寒暑平"，体现了天道公平的思想，古人往往于此时校正度量衡器。

Spring Equinox, the fourth of the Twenty-Four Solar Terms, falls around March 21 on the Gregorian calendar, when the sun reaches celestial longitude 0°. At this time, the sun shines directly on the equator, the southern and northern hemispheres of the earth enjoy an equal share of daytime and nighttime. The spring equinox has "equal daytime and nighttime and thus a fair share of cold and heat", embodying the idea of the heaven being fair to all. Ancient Chinese usually regulated and aligned their timepieces and various measures at this time.

春分三候：一候元鸟至，二候雷乃发声，三候始电。意思是说春分日后，燕子便从南方飞来了，下雨时天空便要打雷并发出闪电。

The three pentads of spring equinox are like this. In the first pentad, swallows start to arrive after their long migratory flight from the south. In the second, when there is rain, it will thunder in roars. In the third, the rain will be accompanied by both thunder and lightning.

二十四节气
The Twenty-Four Solar Terms

春分一候：元鸟至（杨晋 摄）
Swallows Arrive in the First Pentad of Spring Equinox (Photo by Yang Jin)

春分二候：雷乃发声（杨晋 摄）
Thunders Roar in the Second Pentad of Spring Equinox (Photo by Yang Jin)

二十四节气
The Twenty-Four Solar Terms

春分三候：始电（杨晋 摄）
Lightning Seen in the Third Pentad of Spring Equinox (Photo by Yang Jin)

春分时节，进入物候学上真正的春季，大部分地区越冬作物进入春季生长阶段，春播在即，各地先后进入春耕大忙时期。

The actual spring season in the phenological sense arrives around the Spring Equinox. This is a time when most overwinter crops start their phase of spring growth, spring sowing is about to begin, and a busy period of spring tilling and cultivation is launched respectively in most parts of the country.

周代就有春分祭日的仪式。如今民间还有春分竖蛋、吃春菜、赛风筝等习俗。江南等地有粘雀子嘴、犒劳耕牛等习俗，期望农作顺利。客家地区则有在春分时祭祀祖先等习俗。

Ceremonies of making sacrificial offerings to the sun at the Spring Equinox already existed in the Zhou Dynasty. Even now, there are still spring equinox folk customs of standing the egg upright, eating spring vegetables (lettuce, asparagus, or amaranth, depending on the region) and competition in flying kites. In southeast China, there are the customs of feeding sparrows with glutinous rice dumplings in the fields so their beaks will be stuck, and of rewarding the farm cows with glutinous rice balls, to wish for smooth and successful farming. In the Hakka regions, people make sacrificial offerings to their ancestors at Spring Equinox.

春分时节，人体内的阴阳因为天气的变化而上下浮动，容易出现阴阳失衡的情况，因此春分养生关键是调衡阴阳。

At this time, the *yin* and *yang* energy in the human body may fluctuate with the changes in the weather, resulting in an imbalance between the two. Therefore, the key to health building at the spring equinox is the adjustment and balance of *yin* and *yang*.

清明
Freshgreen

清明（郭振毅 摄）
Freshgreen (Photo by Guo Zhenyi)

清明，又名"三月节"或"踏青节"，二十四节气中的第五个节气，公历每年 4 月 4～6 日，太阳到达黄经 15°时。清明是表征春季物候特点的节气，含有天气晴朗、草木繁茂的意思。清明也是中国重要的传统节日。

Freshgreen, otherwise known as "Festival of the Third Month" or "Excursion Festival", the fifth of the Twenty-Four Solar Terms, falls on April 4-6 on the Gregorian calendar, when the sun reaches celestial longitude 15°. This is a solar term that embodies the best phenological features of the spring, including a reference to the clear and bright sky against thriving plants. It is also an important traditional Chinese festival.

　　清明三候：一候桐始华，二候田鼠化为鴽，三候虹始见。意思是说在这个时节先是白桐花开放，接着喜阴的田鼠不见了，全回到了地下的洞中，然后是雨后的天空可以见到彩虹了。

Freshgreen has three pentads. In the first pentad, the princess tree blossoms. In the second, the field mice that prefer the dark and cool are nowhere to be seen, but have all returned underground as if they have transformed into quails and flown away. In the third, rainbows can be observed after the rain.

清明一候：桐始华（杨晋 摄）
Princess Tree Starting to Blossom in the First Pentad of Freshgreen (Photo by Yang Jin)

清明二候：田鼠化为鴽（杨晋 摄）
The Field Mouse Transforming into the Quail in the Second Pentad of Freshgreen (Photo by Yang Jin)

清明三候：虹始见（杨晋 摄）
Rainbow First Observed in the Third Pentad of Freshgreen (Photo by Yang Jin)

清明前后，除东北与西北地区外，中国大部分地区的日平均气温已升到12℃以上，呈现一派生机盎然的春天景象，农业生产进入了一年一度的春耕大忙时节。

Around Freshgreen, except for northeast and northwest China, the average daytime temperature in most parts of the country has risen above 12°C, and thus a spring vista of thriving vitality is presented. Agricultural production enters into the annual busy phase of spring plowing and tilling.

清明民间有扫墓、祭祖、郊游等习俗。四川都江堰地区会举行放水节，用以祭祀修建都江堰水利工程的李冰父子。江苏茅山、顾庄、溱潼等地则有撑会船的习俗。而浙江湖州在这一天有祭拜蚕种的轧蚕花习俗。

Around this solar term, there are folk customs of sweeping the tombs, presenting sacrificial offerings to ancestors, and suburban excursions. In the Dujiangyan area in Sichuan Province, people observe the Water Releasing Festival, to pay respect to Li Bing and his son who built the Dujiangyan Dam Project. Some places in Jiangsu Province, like Maoshan, Guzhuang and Qintong, have the custom of celebrating the Collective Boating Festival. In Huzhou, Zhejiang Province, people observe the custom of presenting sacrificial offerings to silkworm eggs.

清明时期，人体肌肤腠理舒展，五脏六腑因内外清气而润濡，人们宜多到户外运动，如晨练、登山、踏青、郊游等，但要注意防晒、花粉过敏。

During the Freshgreen, the human skin and muscles are relaxed and non-resistant to the environment. The viscera are moistened and refreshed by the clear air indoors and outdoors. People should conduct more outdoor activities, such as morning exercise, mountain climbing, spring excursions, and trips to the suburbs. However, sunburn and allergy to pollen should be guarded against.

谷 雨
Grain Water

谷雨（张争鸣 摄）
Grain Water (Photo by Zhang Zhengming)

二十四节气
The Twenty-Four Solar Terms

谷雨，二十四节气中的第六个节气，公历每年4月19～21日，太阳到达黄经30°时，源自古人"雨生百谷"之说。

Grain Water, the sixth of the Twenty-Four Solar Terms, falls on April 19-21 on the Gregorian calendar, when the sun reaches celestial longitude 30°, deriving its name from the ancient saying that "the rainfall gives rise to a hundred grains".

谷雨三候：一候萍始生，二候鸣鸠拂其羽，三候戴胜降于桑。意思是说谷雨后降雨量增多，浮萍开始生长，接着布谷鸟便开始提醒人们播种了，然后是桑树上开始能见到戴胜鸟。

The three pentads of Grain Water are like this. In the first pentad, after the day of Grain Water, there is more rainfall, and common duckweed starts to grow. In the second, cuckoos peck at their own feathers and chirp to remind people to start sowing seeds. In the third pentad, hoopoes can be seen perched in mulberry trees.

谷雨一候：萍始生（杨晋 摄）
Common Duckweed Starting to Grow in the First Pentad of Grain Water (Photo by Yang Jin)

谷雨二候：鸣鸠拂其羽（杨晋 摄）
Chirping Cuckoo Spreading Its Wings in the Second Pentad of Grain Water (Photo by Yang Jin)

谷雨三候：戴胜降于桑（杨晋 摄）
Hoopoe Perched in the Mulberry Tree in the Third Pentad of Grain Water (Photo by Yang Jin)

二十四节气
The Twenty-Four Solar Terms

谷雨节气的到来意味着寒潮天气基本结束，气温回升加快，有利于谷类农作物的生长。中国大部分地区进入了春种春播的关键时期，是播种移苗、埯瓜点豆的最佳时节。

The arrival of the solar term of Grain Water indicates the ending of the cold and damp weather and the quick rise in temperature, both favorable to the growth of crops of grains. Most of the country now enters the critical period of spring planting and sowing: this is the best timing for sowing seeds, transplanting young seedlings, and dibbling gourd and melon and bean seeds.

谷雨时节，民间有贴禁蝎符驱虫、赏牡丹、吃香椿等习俗。渔民在此时要举行祭海仪式，祈求出海平安、满载而归。黔东南苗族同胞在谷雨时还会举行"爬坡节"，肇兴侗寨有吃乌米、打花脸、播稻种的谷雨节习俗。

During the Grain Water season, people observe such folk customs as having scorpion-repellant talismans against insects, admiring peony flowers, and eating Chinese toon leaves. Fishermen conduct ceremonies of sacrificial offerings to sea gods at this time, to pray for safe navigation and a bumper fishing harvest from the sea. Ethnic Miao people in southeast Guizhou Province celebrate the Slope Climbing Festival at this solar term. Ethnic Dong people in villages in Zhaoxing, Guizhou Province, observe such Grain Water customs as eating black rice (purple glutinous rice), smearing each other's face with cooked black rice, and sowing rice seeds.

谷雨节气降雨增多,空气中的湿度逐渐加大,会让人体由内到外产生不适反应,所以需要针对其气候特点进行调养。谷雨节气后是神经痛的发病期,应注意预防。

During this solar term, the rainfall increases, and the air moisture becomes stronger, so that the human body may experience a series of discomforts inside and outside. Health adjustment and preservation efforts should be managed according to the climatic features. After the Grain Water, there is often an outbreak of neurodynia and prevention should be conducted.

立 夏
Beginning of Summer

立夏（杨晋 摄）
Beginning of Summer (Photo by Yang Jin)

立夏，二十四节气中的第七个节气，公历每年5月5～6日，太阳到达黄经45°时。立夏表示夏季的开始。

Beginning of Summer, the seventh of the Twenty-Four Solar Terms, falls on May 5-6 every year on the Gregorian calendar, when the sun reaches celestial longitude 45°. It indicates the start of summer.

立夏三候：一候蝼蝈鸣，二候蚯蚓出，三候王瓜生。意思是说这一节气中首先可听到蝲蝲蛄（即蝼蛄）在田间的鸣叫声（一说是蛙声），接着大地上便可看到蚯蚓掘土，然后王瓜的蔓藤开始快速攀爬生长。

立夏一候：蝼蝈鸣（杨晋 摄）
Mole Cricket Chirping in the First Pentad of Beginning of Summer (Photo by Yang Jin)

The three pentads of Beginning of Summer are like this. In the first pentad, people can first hear the chirping of mole crickets in the fields (some say it is the croaking of frogs). In the second, earthworms can be seen out of the soil digging earth. In the third, the vines and canes of Trichosanthes cucumeroides start to grow and climb high rapidly.

立夏二候：蚯蚓出（杨晋 摄）
Earthworms Coming Out of the Ground in the Second Pentad of Beginning of Summer (Photo by Yang Jin)

立夏三候：王瓜生（杨晋 摄）
Trichosanthes cucumeroides Growing Fast in the Third Pentad of Beginning of Summer (Photo by Yang Jin)

立夏时节，万物繁茂，气温逐渐升高，炎暑将临，雷雨增多，农作物生长逐渐旺盛，夏收作物进入生长后期，处于产量形成和成熟阶段。水稻栽插以及其他春播作物的管理进入大忙季节。

At Beginning of Summer, all plants thrive and prosper. With the temperature rising and the hot summer approaching, there are more thunderstorms, and farm crops grow more and more vigorously. The crops for summer harvest enter their last phases of growth, a stage of formulation of ultimate yield and also of maturity. A busy season of transplanting rice and attending to the spring-sown crops is started.

立夏日，古代帝王要率文武百官到京城南郊去迎夏，举行迎夏仪式，以表达对丰收的祈求和美好的愿望。古代宫廷里有"立夏日启冰，赐文

武大臣"的习俗。民间则有称人、斗蛋、尝新、煮鼎边做夏、吃光饼等习俗。浙江杭州拱墅区在这天会举办半山立夏节,企盼顺利度夏。

On the day of Beginning of Summer, monarchs in ancient China would head his civilian officials and military officers to welcome Summer in the southern suburbs of the capital city. A summer welcoming ceremony would be hosted, to pray for a bumper harvest and auspiciousness in other areas. There used to be the royal court custom of the emperor having the stored ice taken out on the day of Beginning of Summer and bestowing it to his generals and officials. There were also the folk customs of weighing persons, egg fight, tasting new seasonal produce, eating cauldron-cooked thick soup and eating cakes with a hole in the middle (commemorating Qi Jiguang). In the Gongshu District of Hangzhou, Zhejiang Province, people celebrate the Banshan Beginning of Summer Festival, wishing for a smooth and successful summer time.

立夏时节天气逐渐炎热,万物茂盛,人们的生理状态也发生一定的改变。夏季与心气相通,因此养生首先要养心。

At this solar term, the weather is turning hotter and all living things are thriving. The physiological state of the human body is also undergoing certain changes. Since the summer season is all connected with the *qi* of the heart, health building now prioritizes the preservation of the heart.

小满
Lesser Fullness

小满（杨晋 摄）
Lesser Fullness (Photo by Yang Jin)

小满,二十四节气中的第八个节气,公历每年 5 月 20 ~ 22 日,太阳到达黄经 60°时。小满的含义是夏熟作物的籽粒开始灌浆饱满,但还未成熟,只是小满,还未大满。

Lesser Fullness, the eighth of the Twenty-Four Solar Terms, falls between May 20 and 22, when the sun reaches celestial latitude 60°. Lesser Fullness means that the seeds of summer crops start filling, yet they are not ripe. They have only reached a lesser fullness, not a greater fullness.

小满三候:一候苦菜秀,二候靡草死,三候麦秋至。意思是说小满节气时,苦菜已经枝叶繁茂,而喜阴的一些枝条细软的草类在强烈的阳光下开始枯死,此时麦子开始成熟。

小满一候:苦菜秀(杨晋 摄)
Maror Blossoms in the First Pentad of Lesser Fullness (Photo by Yang Jin)

The three pentads of Lesser Fullness are like this: in the first, maror is seen growing strong; in the second, weak grass dies; in the third, the wheat harvest comes in. In other words, in the solar term of Lesser Fullness, maror prospers and blooms; shade-oriented grass with slender and soft twigs start to die under the strong sunlight; and now wheat starts to ripe.

小满二候：靡草死（杨晋 摄）
Slender Grass Dies in the Second Pentad of Lesser Fullness (Photo by Yang Jin)

二十四节气
The Twenty-Four Solar Terms

小满三候：麦秋至（杨晋 摄）
Wheat Harvest Arrives in the Third Pentad of Lesser Fullness (Photo by Yang Jin)

小满时节，各地气温继续升高，降水继续增多，气温起伏变化大，农事活动即将进入大忙季节，夏收作物已经成熟或接近成熟，春播作物生长旺盛，秋收作物播种在即。

In the solar term of Lesser Fullness, temperature continues to rise all over the country, and precipitation continues to grow. There are great fluctuations in the temperature. People are to enter a very busy farming season. Summer crops are either ripe or are about to ripe. Spring-sown crops grow vigorously and the sowing of autumn crops is about to begin.

俗话说"小满动三车",即水车、油车和丝车,因此民间有小满祭三车的习俗,祈求风调雨顺,日子红火。旧时民间浙江海宁一带还有抢水的习俗。苏浙一带在小满节气期间有一个"祈蚕节",关中地区则有女儿探望父母的看麦梢黄习俗。

A popular saying goes that "three vehicles are set in motion on the day of Lesser Fullness", i.e. the watermill, the oil extractor, and the silk reeling machine. Thus, there is the folk custom of giving sacrifices to the three vehicles, for praying for a bumper harvest and a prosperous life of the year. In the old days, in Haining, Zhejiang, there was also the custom of robbing water. In Jiangsu and Zhejiang, people also celebrated a "Silkworm Festival" while in the Guanzhong region, there was the custom of married daughters visiting their parents and asking after the preparations for the summer harvest.

小满日之后雨量开始增加,湿气较重,加上气温升高,此时是皮肤病的高发期,要秉承"未病先防"的观点,加强预防。此时人的情绪波动也较大,平时要注意控制自己的情绪。

After the Day of Lesser Fullness, rainfall starts to increase, moisture in the air grows, and temperature rises. Thus it is a season of greater outbreak of skin diseases. The idea of "prevention of diseases when one is healthy" should be subscribed to, and prevention should be on the agenda. People are subject to greater emotional fluctuations around this time. One should restrain one's own temper and mood.

芒种
Grain in Ear

芒种（郭振毅 摄）
Grain in Ear (Photo by Guo Zhenyi)

芒种,二十四节气中的第九个节气,公历每年 6 月 6 日前后,太阳到达黄经 75°时。此时大麦、小麦等有芒作物种子已经成熟,抢收十分急迫。晚谷、黍、稷等夏播作物也正是播种最忙的季节,故又称"忙种"。

Grain in Ear, the ninth of the Twenty-Four Solar Terms, falls around June 6 on the Gregorian calendar, when the sun reaches celestial longitude 75°. At this time, awn crops such as barley and wheat are ripe, and harvest is urgent. It is also the busiest sowing season for such summer-sown crops as late corn, glutinous millet, and millet.

芒种三候:一候螳螂生,二候䴗始鸣,三候反舌无声。意思是说在这一节气中,螳螂在上一年深秋产的卵,因感受到阴气初生而破壳生出小螳螂;喜阴的伯劳鸟开始在枝头出现,并且感阴而鸣。与此相反,能够学习其他鸟叫的反舌鸟,却因感应到了阴气的出现而停止了鸣叫。

The three pentads of Grain in Ear are like this. In the first pentad, the mantis is hatched. In the second, the butcher bird starts to chirp. In the third, the mockingbird becomes silent. During this solar term, the mantis eggs produced in late autumn of the previous year have felt the rise of the *yin* energy and thus baby mantises grow out of their shells. The butcher bird that loves shady places starts to appear on tree branches and chirp whenever it feels the *yin* energy. On the other hand, the mockingbird that can imitate the songs of other birds stops chirping altogether as it feels the rise the *yin* energy.

二十四节气
The Twenty-Four Solar Terms

芒种一候：螳螂生（杨晋 摄）
Mantis Hatched in the First Pentad of Grain in Ear (Photo by Yang Jin)

芒种二候：鵙始鸣（杨晋 摄）
The Butcher Bird Starting to Chirp in the Second Pentad of Grain in Ear (Photo by Yang Jin)

二十四节气
The Twenty-Four Solar Terms

芒种三候：反舌无声（杨晋 摄）
Mockingbirds Stopping Singing in the Third Pentad of Grain in Ear (Photo by Yang Jin)

芒种一到，夏熟作物要收获，夏播秋收作物要下地，春种的庄稼要管理，收、种、管交叉，是一年中最忙的季节。

As soon as we enter the solar term of Grain in Ear, the summer harvest crops need to be gathered in and the summer sown crops need to be sown or planted. Spring sown crops must be taken care of. It is really the busiest season of the year with harvesting, sowing and nurturing going on at the same time.

芒种时节百花凋落，因此民间有饯送花神习俗，感谢花神，期盼来年再次相会。皖南地区会举行安苗祭祀活动，祈求农业丰收。贵州东

南部地区则有插秧时打泥巴仗的习俗，缓解劳动的辛劳。此外芒种节气还是青梅成熟的时候，因此古时还有煮梅的习俗。

In the solar term of Grain in Ear, the various flowers fade. Thus, around this time, people have the folk custom of giving a farewell party to the deity of the flowers, to thank him for his favors and to wish for a happy reunion in the following year. In south Anhui Province, people have the sacrificial events for tranquilizing the seedlings and wishing for a bumper agricultural harvest. In southeast Guizhou Province, people have the custom of clay mud fight when they transplant rice seedlings, in order to ease their fatigue from labor. Meanwhile, this is also the season for the ripening of the greengage, and therefore in ancient times people had the custom of boiling plums for food.

芒种时期，天气炎热，已经进入真正的夏季。此时要根据季节的气候特征，注意精神调养，清除体内毒素，保持充足的睡眠，为即将到来的暑气做好准备。

In the Grain in Ear period, it is hot, as we are already in real summer. One should, due to the climatic features of the season, regulate one's mood, eliminate the poisonous elements in the body, and have sufficient hours of sleep, making ample preparations for the soon-to-arrive summer heat.

夏 至
Summer Solstice

夏至（彭徐蒙 摄）
Summer Solstice (Photo by Peng Xumeng)

二十四节气
The Twenty-Four Solar Terms

夏至，二十四节气中的第十个节气，公历每年 6 月 21～22 日，太阳到达黄经 90°时。它是二十四节气中最早被确定的节气之一。夏至这天，太阳直射地面的位置到达一年的最北端，北半球的白昼达到最长。

Summer Solstice, the tenth of the Twenty-Four Solar Terms, falls on June 21 or 22 on the Gregorian calendar, when the sun reaches celestial longitude 90°. It is one of the solar terms established the earliest among all the twenty-four. On the day of Summer Solstice, the sun shines directly on a position on the earth that is the northernmost in the year, and daylight on the northern hemisphere is the longest for the year.

夏至三候：一候鹿角解，二候蜩始鸣，三候半夏生。意思是说夏至日阴气生而阳气始衰，所以阳性的鹿角便开始脱落；雄性的知了在夏至后因感阴气之生便鼓翼而鸣；半夏是一种喜阴的药草，因在仲夏的沼泽地或水田中出生所以得名。

The three pentads are like this. In the first, deer shed their horns; in the second, cicadas start to buzz; and in the third, tuber pinellia appears. In other words, on the day of Summer Solstice, the negative *qi* (*yin* energy) starts to grow and the positive *qi* (*yang* energy) starts to decline, and thus the horns of deer which fall into the category of positive energy start to be shed; after the day of Summer Solstice, male cicadas feel the advent of the negative *qi* and sound their wings to buzz; tuber pinellia is a medicinal herb that favors shady places, and are thus taking their roots in marshlands or rice paddies in mid-summer.

二十四节气
The Twenty-Four Solar Terms

夏至一候：鹿角解（杨晋 摄）
Deer Shed Their Horns in the First Pentad of Summer Solstice (Photo by Yang Jin)

夏至二候：蜩始鸣（杨晋 摄）
Cicadas Start to Buzz in the Second Pentad of Summer Solstice (Photo by Yang Jin)

夏至三候：半夏生（杨晋 摄）
Tuber Pinellia Appears in the Third Pentad of Summer Solstice (Photo by Yang Jin)

夏至时节，中国大部分地区气温较高，日照充足，作物生长最为旺盛，降水较多。南北各地水稻插秧都将结束，冬小麦大多已成熟，春播作物已由苗期进入生长中期，夏播作物还处于苗期。

In Summer Solstice, most regions in China have high temperatures with plenty of sunshine, and crops grow most vibrantly with abundant rain. Rice transplanting in both north and south China is about to conclude. Most winter wheat has ripened. Spring-sown crops have transformed from the seedling stage into the growing period, while the summer-sown crops are still in the seedling stage.

夏至时值麦收,自古以来有在此时庆祝丰收、祭祀祖先之俗。为了消夏避伏,妇女们有在此时互相赠送折扇、脂粉等的习俗。此外民间还有吃面条、麦粽、夏至饼和戴枣花等习俗。

Summer Solstice is a time of wheat harvest, a time to celebrate harvests and to pay homage to ancestors since ancient times. In order to mitigate the effects of summer and its intense heat, women sometimes present each other with folded fans and cosmetics as a custom at this time. Besides, people also have the custom of eating noodles, wheat dumplings, Summer Solstice cakes, and wearing flowers of date.

夏至是阳气较旺的时节,养生要顺应夏季阳盛于外的特点,注意保护阳气。炎热的气候,会减弱人体的消化功能,要从生活起居和饮食上保护好消化功能。

Summer solstice is a time with vibrant *yang* energy. Health building should conform to the characteristic of *yang* being strong on the external in summer and one should protect the *yang* energy. The climate of intense heat weakens the body's digestive functions, and thus one must protect the digestive functions with respect to daily habits and diet.

小 暑
Lesser Heat

小暑（杨晋 摄）
Lesser Heat (Photo by Yang Jin)

小暑，二十四节气中的第十一个节气，公历每年 7 月 7 ～ 8 日，太阳到达黄经 105°时。小暑期间，还不是一年中最热的时候，故称为小暑。

Lesser Heat, the eleventh of the Twenty-Four Solar Terms, falls on July 7 or 8 each year on the Gregorian calendar, when the sun reaches celestial longitude 105°. This period is not the hottest in the year, and thus it is referred to as Lesser Heat.

小暑三候：一候温风至，二候蟋蟀居壁，三候鹰始击。意思是说小暑时节大地上所有的风中都带着热浪；由于炎热，蟋蟀离开了田野，到庭院的墙角下以避暑热；老鹰因地面气温太高而在清凉的高空中活动。

小暑一候：温风至（杨晋 摄）
Warm Breezes Arrive in the First Pentad of Lesser Heat (Photo by Yang Jin)

The three pentads of Lesser Heat are like this. In the first pentad, the warm breeze arrives. In the second, crickets are found in homes. In the third, hawks start to fly high. In other words, winds blowing on the earth in this solar term carry heat waves with them. Due to the rising heat, crickets leave the fields and seek refuge in the corners of the walls in courtyards. Hawks start to fly high where the air is cool, as the surface of the earth is too hot for them.

小暑二候：蟋蟀居壁（杨晋 摄）
Crickets Are Found in Courtyards in the Second Pentad of Lesser Heat (Photo by Yang Jin)

小暑三候：鹰始击（杨晋 摄）
Hawks Start to Fly High in the Third Pentad of Lesser Heat (Photo by Yang Jin)

小暑前后，除东北与西北地区收割冬、春小麦等作物外，全国大部分地区的夏秋作物，得益于此时较高的温度、丰沛的雨水、充足的光照，进入生长最为旺盛的时期，农田管理进入较为繁忙的时期。

Around Lesser Heat, in northeast and northwest China, crops like winter and spring wheat are harvested. But the summer and autumn crops in most parts of China enter a phase of most vibrant growth, benefiting from the high temperature, plentiful rainfall and sufficient sunlight of this period. It is now a busy period for attendance to farm fields.

小暑时节正值伏日，民间一般有吃饺子、面等习俗。山东部分地区有吃生黄瓜和煮鸡蛋来治苦夏的习俗，江苏徐州地区有入伏吃羊肉俗称"吃伏羊"的习俗，南方部分地区则有小暑吃蜜汁藕的习惯。

Lesser Heat is the dog days of the summer. Folk customs include eating dumplings (*jiaozi*) and noodles. In some parts of Shandong Province, such customs as eating raw cucumbers and boiled eggs exist for confronting summer loss of appetite and weight. In Xuzhou, Jiangsu Province, the custom is eating mutton as soon as it enters the dog days, popularly known as "eating mutton in the dog days". In some parts of south China, people have the habit of eating lotus roots dipped in honey dews during this solar term.

小暑时节，标志着夏天最热的时候已经来临。这时的高温气候常使人感觉心烦不安，疲倦乏力，食欲下降，此时要顺应节气变换，增进食欲，保证必要的营养。

Lesser Heat marks the advent of the hottest time of summer. High temperatures around this time often render people restless, fatigued, weak, and experience a loss of appetite. People need to conform to the shift of the solar term, inspire the appetite and supply necessary nutrients.

大 暑
Greater Heat

大暑（杨晋 摄）
Greater Heat (Photo by Yang Jin)

二十四节气
The Twenty-Four Solar Terms

大暑，二十四节气中的第十二个节气，公历每年 7 月 23～24 日，太阳到达黄经 120°时。大暑是一年中最热的节气。

Greater Heat, the twelfth of the Twenty-Four Solar Terms, falls on July 23-24 each year on the Gregorian calendar, when the sun reaches celestial longitude 120°. This is the hottest solar term in the year.

大暑三候：一候腐草为萤，二候土润溽暑，三候大雨时行。意思是说大暑时会出现萤火虫；接着天气开始变得闷热，土地也很潮湿；然后时常有大的雷雨会出现，这大雨使暑湿减弱，天气开始向立秋过渡。

The three pentads are like this. In the first, fireflies abound in decayed grass; in the second, earth is moist in this damp and sultry weather; in the third, heavy rainfalls are frequent. In other words, in this solar term, fireflies can often be seen. Then the weather starts to become sultry, and earth is moist and damp. Afterwards, there are often thunderstorms, which reduce the intense heat and dampness. The weather begins a transition to the Beginning of Autumn.

大暑一候：腐草为萤（杨晋 摄）
Fireflies in the Decayed Grasses in the First Pentad of Greater Heat (Photo by Yang Jin)

大暑二候：土润溽暑（杨晋 摄）
Earth Moistened in Damp Heat in the Second Pentad of Greater Heat (Photo by Yang Jin)

二十四节气
The Twenty-Four Solar Terms

大暑三候：大雨时行（杨晋 摄）
Frequent Heavy Rainfalls in the Third Pentad of Greater Heat (Photo by Yang Jin)

大暑节气正值"中伏"前后，全国大部分地区进入一年中最热的时期，也是喜温作物生长最快的时期，但旱、涝、台风等自然灾害发生频繁，抢收抢种、抗旱排涝防台风和田间管理等任务最重。

The solar term of Greater Heat is the middle phase of the dog days. Most regions of China are now experiencing the hottest weather in the year. This is also a period of fastest growth for warm-season crops. However, natural disasters such as droughts, floods and typhoons are frequent occurrences. People have to carry out heavy tasks of hasty harvests, hasty planting, fight with droughts and floods, precautions against typhoons, and crop attendance in farm fields.

大暑时节，浙江沿海地区，特别是台州渔村有送"大暑船"的习俗，其意送暑保平安。山东南部地区有喝羊汤的习俗。浙江台州椒江人有吃

姜汁调蛋的风俗。福建莆田人有吃荔枝、羊肉和米糟的习俗，叫作"过大暑"。广东很多地方则有"吃仙草"的习俗。

During this solar term, in coastal areas of Zhejiang Province, especially in villages of Taizhou, people have the custom of presenting gifts of "Greater Heat Boats", signifying sending off heat and praying for peace and safety. In the south of Shandong Province, people have the custom of eating mutton soup; in Jiaojiang District of Taizhou, Zhejiang Province, people eat a custard-like egg soup in a sauce of ginger, sugar, and glutinous rice wine. People in Putian, Fujian Province have the custom of eating lychees, mutton and rice dregs, known as "Celebrating Greater Heat". In many places in Guangdong Province, on the other hand, people have the custom of "eating Chinese jelly grass".

大暑是全年气温最高、阳气最盛的时节，此时要预防中暑。同时在养生保健中常有"冬病夏治"的说法，故对于那些每逢冬季发作的慢性疾病，如慢性支气管炎、肺气肿、支气管哮喘、腹泻、风湿痹症等阳虚症，是最佳的治疗时机。

Greater Heat is the season when the temperature is the highest and the *yang qi* energy is the strongest in the year. This is also the time to take precautions against heat stroke. In terms of health preservation, there is the saying that winter diseases had better be treated in summer, and thus this is best timing for *yang* deficiency syndromes such as chronic bronchitis, emphysema, diarrhoea and rheumatism.

立 秋
Beginning of Autumn

立秋（杨晋 摄）
Beginning of Autumn (Photo by Yang Jin)

立秋，二十四节气中的第十三个节气，公历每年 8 月 7~8 日，太阳到达黄经 135° 时。立秋不仅预示着炎热的夏天即将过去，秋天即将来临，也表示草木开始结果孕子，收获季节到了。

Beginning of Autumn, the thirteenth of the Twenty-Four Solar Terms, falls on August 7-8 on the Gregorian calendar, when the sun reaches celestial latitude 135°. It means the imminent departure of the hot summer season and the approach of autumn. It also indicates that grass and trees are starting to bear fruit and seeds, and that the season of harvest is coming.

立秋三候：一候凉风至，二候白露降，三候寒蝉鸣。意思是说立秋过后，刮风时人们会感觉到凉爽，此时的风已不同于夏天中的热风，接着，大地上早晨会有雾气产生，并且秋天感阴而鸣的寒蝉也开始鸣叫。

The three pentads of Beginning of Autumn are like this. In the first pentad, cool breezes arrive. In the second, white dews fall. In the third, cicadas buzz in the cooler weather. In other words, after the day of Beginning of Autumn, the breezes that blow convey a tinge of coolness, unlike the wind of heat in summer. Soon, fog starts to arise in the morning on the ground, and cicadas that make noises when feeling the *yin* energy start to buzz.

二十四节气
The Twenty-Four Solar Terms

立秋一候：凉风至（杨晋 摄）
Cool Breezes Arrive in the First Pentad of Beginning of Autumn (Photo by Yang Jin)

立秋二候：白露降（杨晋 摄）
White Dews Descend in the Second Pentad of Beginning of Autumn (Photo by Yang Jin)

立秋三候：寒蝉鸣（杨晋 摄）
Cicadas Buzz at Cooler Weather in the Third Pentad of Beginning of Autumn (Photo by Yang Jin)

　　立秋前后中国大部分地区气温仍然较高，各种农作物生长旺盛，中稻开花结实，单晚圆秆，大豆结荚，玉米抽雄吐丝，棉花结铃，甘薯薯块迅速膨大，此时对水分要求都很迫切，要注意及时浇灌。此时也是作物病虫害集中的时期，要加强预测预报和防治。

Around Beginning of Autumn, the temperature is still very high in most parts of China. Agricultural crops are growing strong. Medium rice is blooming and bearing fruit. The strengthening of serotinous single-crop rice is being completed. Soybeans are podding. Corn is tasselling and earing. Cotton bolls begin to fill. The tube of the sweet potato is swelling fast. They all require moisture and water urgently and thus watering should be provided in a timely manner. This is also a period when crop diseases and insect pests break out intensively and thus forecasts, early warning, prevention and treatment need to be strengthened.

古代立秋时节有天子到西郊迎秋祭祀的习俗，宋代立秋之日，男女都戴楸叶，以应时序。如今湖南花垣等地，苗族在这一天会举行隆重的赶秋节。此外立秋还有啃秋、陈冰瓜、蒸茄脯、煎香薷饮、尝新、奠祖等习俗。

In ancient times, the custom at Beginning of Autumn was for the Son of Heaven to usher in the Autumn and make sacrificial offerings in western suburbs of the capital city. On the day of Beginning of Autumn in the Song Dynasty, both men and women wore Manchurian catalpa leaves to acknowledge and accommodate the season. In some regions today, such as Huayuan, Hunan Province, ethnic Miao people hold solemn ceremonies of amusement and celebrations to welcome the autumn. Moreover, in this solar term, there are also customs like eating water melons, displaying iced water melons, steaming eggplant preserves, having fried elsholtzia drinks, tasting seasonal vegetables, and offering sacrifices to ancestors.

立秋日之后虽然一时暑气难消，还有"秋老虎"的余威，但总的趋势是天气逐渐凉爽。此时自然界的阳气变化也从"长"的状态转向"收"的状态。此时需注意精神调养，饮食上宜补养脾胃，多吃些清热祛燥的食物。

After the Day of Beginning of Autumn, although summer heat does not take its leave easily, it still inflicts a lingering aggressiveness of the "autumn tiger", but on the whole, it is turning cooler. Around this time, the *yang* energy in the nature switched from advancement to retreat. At this time, one needs to make mood adjustment, and in regimen, it does one good to boost the spleen and stomach by eating more food that clears heat and dispels dryness.

处 暑
End of Heat

处暑（杨晋 摄）
End of Heat (Photo by Yang Jin)

二十四节气
The Twenty-Four Solar Terms

处暑，二十四节气中的第十四个节气，公历每年 8 月 23 日前后，太阳到达黄经 150°时。处暑意味着炎热的夏天即将过去。

End of Heat, the fourteenth of the Twenty-Four Solar Terms, falls around August 23 on the Gregorian calendar each year, when the sun reaches celestial longitude 150°. It signifies that the hot summer is about to end.

处暑三候：一候鹰乃祭鸟，二候天地始肃，三候禾乃登。此节气中老鹰开始大量捕猎鸟类，天地间万物开始凋零，黍、稷、稻、粱类农作物开始成熟。

The three pentads of End of Heat are like this. In the first pentad, hawks start to make bird offerings. In the second, heaven and earth start to become stern and gloomy. In the third, millets ripen. In other words, hawks start to prey on other birds in large numbers; all living things start to wither; crops like millet, glutinous millet, rice and sorghum start to ripen.

二十四节气
The Twenty-Four Solar Terms

处暑一候：鹰乃祭鸟（杨晋 摄）
Hawks Start to Prey on Birds in the First Pentad of End of Heat (Photo by Yang Jin)

处暑二候：天地始肃（杨晋 摄）
Heaven and Earth Start to Be Stern and Gloomy in the Second Pentad of End of Heat (Photo by Yang Jin)

处暑三候：禾乃登（杨晋 摄）
Millets Start to Ripen in the Third Pentad of End of Heat (Photo by Yang Jin)

  处暑节气以后，全国大部分地区的气温逐渐降低，降水逐渐减少，昼夜温差较大，有利于作物体内干物质的形成和积累，因此，处暑以后，各种庄稼成熟得格外快。中国南方大部分地区这时也正是收获中稻的大忙时节。

After the solar term of End of Heat, in most regions of China, the temperature gradually drops, and rainfall becomes less. The greater gap in temperature between day and night is conducive to the formation and accumulation of the dry elements inside crops. Therefore, after End of Heat, the various crops ripen at a fast pace. In most areas of southern China, this is the busy season for harvesting medium rice.

浙江沿海在处暑时要举行隆重的开渔节,欢送渔民开船出海。民间还有处暑吃鸭子的习俗。

On coastal areas of Zhejiang Province, people hold a very ceremonial festival of launch of fishing at End of Heat, wishing the fishermen well when they set their sails on their sea-faring fishing trips. There is also the folk custom of eating duck at this solar term.

《黄帝内经》认为,处暑后阳气减弱、阴气增长,人体内阴阳之气的盛衰也随之转换。中医认为,处暑占有"暑"和"燥"两种外邪,肺与秋季相应,因此要注意养肺。

According to *Huang Di Nei Jing* (Yellow Emperor's Inner Canon), after End of Heat, the *yang* energy declines and the *yin* energy grows. The prosperity and decline of *yin* and *yang* energy in the human body also start to switch. In traditional Chinese medicine, End of Heat is endowed with exogenous evils of both "heat" and "dryness". The lungs correspond to autumn and so one needs to take special care of them from now on.

白 露
White Dew

白露(赵敏 摄)
White Dew (Photo by Zhao Min)

二十四节气
The Twenty-Four Solar Terms

白露,二十四节气中的第十五个节气,公历每年9月7~9日,太阳到达黄经165°时。白露时节,天气逐渐转凉,清晨时分会在地面和叶子上发现有许多露珠,这是因夜晚水汽凝结在上面,故名。

White Dew, the fifteenth of the Twenty-Four Solar Terms, falls on September 7-9, when the sun reaches celestial longitude 165°. During this solar term, it gradually turns chilly; many drops of dew are found on the ground and on leaves in the morning, as vapors coagulate there in the night, thus the name White Dew.

白露三候:一候鸿雁来,二候元鸟归,三候群鸟养羞。意思是说时值白露节气,鸿雁与燕子等候鸟准备南飞避寒,百鸟开始贮存干果粮食以备过冬。

The three pentads of White Dew are like this. In the first pentad, swan geese arrive; in the second, swallows return; and in the third, all birds of feather stock food. In other words, in this solar term, migrant birds like swan geese and swallows are ready to fly to the south to stay away from cold, and various birds and fowl start their stock collections for the winter.

二十四节气
The Twenty-Four Solar Terms

白露一候：鸿雁来（杨晋 摄）
Swan Geese Arrive in the First Pentad of White Dew (Photo by Yang Jin)

白露二候：元鸟归（杨晋 摄）
Swallows Return in the Second Pentad of White Dew (Photo by Yang Jin)

白露三候：群鸟养羞（杨晋 摄）
All Birds of Feather Stock Food in the Third Pentad of White Dew (Photo by Yang Jin)

　　白露时节，全国普遍进入金秋季节，秋高气爽，农业生产方面此时也是抢收庄稼和秋播的大好时节。

In the solar term of White Dew, the whole country enters the harvest season. The clear and crisp autumn days are the best agricultural season for autumnal harvest and sowing within a short time frame.

　　白露节气这一天，福州民间有吃龙眼补身体的习俗，浙江温州等地民间有采集"十样白"煨乌骨白毛鸡（或鸭子）过白露节的风俗，苏浙一带有白露酿酒的习俗。此外，太湖地区渔民还会举行祭禹王的香会。

On the Day of White Dew, in Fuzhou, Fujian Province, people have the folk custom of eating longan fruit to nourish and stimulate one's health. In Wenzhou, Zhejiang Province, people have the folk custom of gathering "ten white foodstuffs" for stewing black-boned white-feathered chicken (or ducks). In Jiangsu and Zhejiang Provinces, people have the custom of brewing wine with white dew. In the Taihu Lake area, fishermen give sacrificial offerings to Emperor Yu at a pilgrim ceremony.

白露时节，天气明显转凉，最常见的疾病就是感冒、过敏性鼻炎、咽炎和秋季腹泻等，因此应增强免疫力，从生活的各个方面预防秋季易发病。

During the solar term of White Dew, it is conspicuously turning chilly. The most common diseases include cold, allergic rhinitis, pharyngitis, and autumnal diarrhea. Thus, to prevent autumnal diseases in various respects of life, one needs to enhance immunity.

秋分

Autumn Equinox

秋分（唐寒飞 摄）
Autumn Equinox (Photo by Tang Hanfei)

二十四节气
The Twenty-Four Solar Terms

秋分，二十四节气中的第十六个节气，公历每年9月22～23日，太阳到达黄经180°时。此时太阳直射地球赤道，昼夜均分，全球无极昼极夜现象。秋分之后白天逐渐变短，黑夜变长，气温逐日下降，逐渐步入深秋季节。

Autumn Equinox, the sixteenth of the Twenty-Four Solar Terms, falls on September 22-23 each year on the Gregorian calendar, when the sun reaches celestial longitude 180°. At this time, the sun shines directly upon the equator of the earth; daytime and nighttime are equal; there is no phenomenon of polar day or polar night. After Autumn Equinox, daytime starts to get shorter and nighttime becomes longer. The temperature drops day by day and it is getting into the season of deep autumn gradually.

秋分三候：一候雷始收声，二候蛰虫坏户，三候水始涸。意思是说秋分后阴气开始旺盛，所以不再打雷了；五日后由于天气变冷，蛰居的小虫开始藏入穴中，并且用细土将洞口封起来以防寒气侵入；再五日降雨量开始减少，由于天气干燥，水汽蒸发快，所以湖泊与河流中的水量变少，一些沼泽及水洼处便处于干涸之中。

The three pentads of Autumn Equinox are like this. In the first pentad, thunders hide themselves. In the second pentad, worms start to build their houses for hibernation. In the third, river water starts to dry up. In other words, after Autumn Equinox, *yin* energy starts to be strong and thus it no longer thunders. Five days later, due to the colder weather, some worms that hibernate start to hide themselves in caves and seal the openings with fine earth to prevent cold air from invasion. Five more days later, due to the decrease in rainfall, and the weather being dry, and moisture evaporating fast, the water volume in lakes and rivers becomes less. Some marshes and loblollies dry up.

二十四节气
The Twenty-Four Solar Terms

秋分一候：雷始收声（杨晋 摄）
Thunders Become Silent in the First Pentad of Autumn Equinox (Photo by Yang Jin)

秋分二候：蛰虫坏户（杨晋 摄）
Worms Seal Their Caves for Hibernation in the Second Pentad of Autumn Equinox (Photo by Yang Jin)

秋分三候：水始涸（杨晋 摄）
Water Dries Up in the Third Pentad of Autumn Equinox (Photo by Yang Jin)

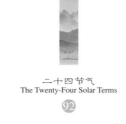

秋分时节降温快，使得秋收、秋耕、秋种的"三秋"大忙显得格外紧张。

During this solar term, the temperature drops quickly, and the autumn season of three busy activities—harvest, plowing and sowing—is especially tight-scheduled.

古人在秋分时会举行祭月仪式。岭南地区在秋分节气有吃秋菜的习俗。农村还有用汤圆粘雀子嘴，防止雀子偷吃庄稼等习俗。

Ancient Chinese held ceremonies to make sacrificial offerings to the moon at Autumn Equinox. In Lingnan area, people had the custom of eating autumn vegetables. In the countryside, people stuck beaks of sparrows with glutinous rice dumplings to stop them from stealing and damaging crops.

秋分节气已经真正进入到秋季，作为昼夜时间相等的节气，人们在养生中也应本着阴阳平衡的规律，使机体保持"阴平阳秘"的原则。

In this solar term, it gets into the real autumn season, and daytime and nighttime are equal to each other. In health building, people should obey the rule of balance between *yin* and *yang*, so that the human body can also subscribe to the principle of "*yin* is peacefully calm and *yang* holds onto itself ".

寒 露
Cold Dew

寒露（杨晋 摄）
Cold Dew (Photo by Yang Jin)

寒露，二十四节气中的第十七个节气，公历每年10月8～9日，太阳到达黄经195°时。寒露的意思是气温比白露时更低，地面的露水更冷，快要凝结成霜了。

Cold Dew, the seventeenth of the Twenty-Four Solar Terms, falls on October 8-9 each year on the Gregorian calendar, when the sun reaches celestial longitude 195°. "Cold Dew" means that the temperature is even lower than the time of "White Dew", and that dewdrops on the ground are so cold that they are about to condense into frost.

寒露三候：一候鸿雁来宾，二候雀入大水为蛤，三候菊有黄华。意思是说此节气中鸿雁开始南迁；深秋天寒，雀鸟都不见了，古人看到海边突然出现很多蛤蜊，并且贝壳的条纹及颜色与雀鸟很相似，便以为是雀鸟变成的；此时菊花已普遍开放。

These are the three pentads of Cold Dew. In the first pentad, swan geese start their migration. In the second, birds disappear into the waters to change into clams. In the third, chrysanthemums are seen blooming. In other words, swan geese start their migration to the south; in the depth of autumn, birds have disappeared while on the sea coast ancient Chinese suddenly saw many clams instead; with the stripes on clam shells look similar to the patterns on bird feathers, they had the illusion that birds transformed into clams; at this time, chrysanthemums are seen in full blossom everywhere.

二十四节气
The Twenty-Four Solar Terms

寒露一候：鸿雁来宾（杨晋 摄）
Swan Geese Start Migration to the South in the First Pentad of Cold Dew (Photo by Yang Jin)

寒露二候：雀入大水为蛤（杨晋 摄）
Birds Disappear Into the Sea to Become Clams in the Second Pentad of Cold Dew (Photo by Yang Jin)

寒露三候：菊有黄华（杨晋 摄）
Chrysanthemums in Full Yellow Blossoms in the Third Pentad of Cold Dew (Photo by Yang Jin)

寒露时节，江淮及江南的单季晚稻即将成熟，双季晚稻正在灌浆，要注意间歇灌溉，保持田间湿润，同时要注意防御"寒露风"的危害。华北地区要抓紧播种小麦。长江流域适宜直播油菜，淮河以南的绿肥播种要抓紧扫尾，棉花和甘薯都要抓紧采收。

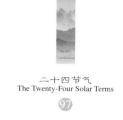

During this solar term, the single-season late rice in the areas between Yangtze River and Huaihe River, and also to the south of the middle and lower reaches of the Yangtze River is ripening. Double-season rice is grouting. Care should be taken to do irrigation at intervals, to keep the fields moist and to guard against the hazards of "Cold Dew wind". Wheat sowing should be hurried in northern China. Direct seeding and sowing of rape is suitable for the Yangtze River basin. Green manure sowing should be done as soon as possible to the south of the Huaihe River. Harvesting of cotton and sweet potatoes should be completed within a very tight time frame.

寒露之时，在一些地方有秋钓边、吃芝麻、赏红叶、吃螃蟹等习俗。

At Cold Dew, there are customs of fishing at shallow places, eating sesames, admiring red maple leaves and eating crabs.

寒露以后，随着气温的不断下降，最应警惕心脑血管疾病，另外，脑卒中、老年慢性支气管炎、哮喘病、肺炎等也容易复发，要采取综合措施，积极预防，合理安排日常起居生活。

After Cold Dew, with the ongoing drop in temperature, cardiovascular and cerebrovascular diseases in the heart and brain should be guarded against. Besides, cerebral apoplexy, geriatric chronic bronchitis, asthma, and pneumonia relapse easily. Comprehensive measures should be taken for active prevention. Appropriate arrangements should be made for daily life.

霜降
First Frost

霜降(冯承庆 摄)
First Frost (Photo by Feng Chengqing)

二十四节气
The Twenty-Four Solar Terms

　　霜降，二十四节气中的第十八个节气，公历每年 10 月 23～24 日，太阳到达黄经 210°时。此时，中国黄河流域开始出现白霜。

First Frost, the eighteenth of the Twenty-Four Solar Terms, falls on October 23-24 each year on the Gregorian calendar, when the sun reaches celestial longitude 210°. At this time, white frost starts to appear in the Yellow River basin in China.

　　霜降三候：一候豺乃祭兽，二候草木黄落，三候蛰虫咸俯。意思是说豺狼开始捕获猎物，做出祭兽的样子；大地上的树叶枯黄掉落；蛰虫也全在洞中不动不食，垂下头来进入冬眠状态。

霜降一候：豺乃祭兽（杨晋　摄）
Jackals Pray to Their Prey in the First Pentad of First Frost (Photo by Yang Jin)

The three pentads of First Frost are like this: in the first pentad, jackals pray to prey; in the second pentad, grass and trees turn yellow and leaves fall; in the third, ground beetles hibernate. In other words, jackals and wolves start to prey on other animals as if they are doing a praying gesture; tree leaves on the ground wither and yellow and fall; ground beetles all hide themselves in caves with no eating or movement and hang their heads low in hibernation.

霜降二候：草木黄落（杨晋 摄）
Grass and Trees Turn Yellow, and Leaves Fall in the Second Pentad of First Frost (Photo by Yang Jin)

二十四节气
The Twenty-Four Solar Terms

霜降三候：蜇虫咸俯（杨晋 摄）
All Ground Beetles Go into Caves and Hang Their Heads in the Third Pentad of First Frost (Photo by Yang Jin)

　　霜降时节，正是农业生产上处于季节转换、年度更替之时，此时，北方大部分地区已在秋收扫尾，在南方却是"三秋"大忙季节。

The solar term of First Frost is at a time when agriculture and farming go through seasonal shifts and annual alteration. Around this time, in most parts of northern China, autumn harvest is near its end, while in the south, it is an exceedingly busy season of autumn harvest, plowing and planting.

霜降时节，广西壮族自治区天等县向都镇等地会举行盛大的霜降节活动，民间还有吃柿子等习俗。

In the First Frost season, spectacular First Frost Festival events are held in such places as Xiangdu Town, Tiandeng County, Guangxi Zhuang Autonomous Region. There is also the popular custom of eating persimmons.

养生方面，民间有谚语"一年补透透，不如补霜降"，霜降时节首先要重视保暖，其次要防秋燥，运动量可适当加大。

In terms of health building, as a folk proverb goes, "nutrition enhancement for a whole year is no better than once at the First Frost". Around this time, people should make a point of keeping warm. They should then guard against autumn dryness diseases. They are advised to increase their amount of exercise.

立 冬
Beginning of Winter

立冬（贺敬华 摄）
Beginning of Winter (Photo by He Jinghua)

立冬,二十四节气中的第十九个节气,公历每年 11 月 7 ~ 8 日,太阳到达黄经 225°时。立冬不仅意味着冬季的开始,还有万物收藏、规避寒冷的意思。

Beginning of Winter, the nineteenth of the Twenty-Four Solar Terms, falls on November 7-8 each year on the Gregorian calendar, when the sun reaches celestial longitude 225°. This solar term does not only mean the beginning of winter, but also the collection and stocking of everything and avoidance and prevention of cold.

立冬三候:一候水始冰,二候地始冻,三候雉入大水为蜃。意思是说从此节气开始,水已经能结成冰;土地也开始冻结;野鸡一类的大鸟便不多见了,而海边却可以看到外壳与野鸡的线条及颜色相似的大蛤,所以古人认为雉到立冬后便变成大蛤了。

Here are the three pentads of Beginning of Winter. In the first pentad, surface water starts icing; in the second pentad, the earth starts to freeze; in the third, pheasants enter the seas and become clams. In other words, beginning from this solar term, water can freeze to become ice; the earth starts to freeze; big birds such as the pheasant are nowhere to be seen, but on the seashore, people can observe big clams whose stripes and colors are similar to those of the pheasants, and thus ancient Chinese people believed that pheasants turn into clams after the Beginning of Winter.

二十四节气
The Twenty-Four Solar Terms

立冬一候：水始冰（杨晋 摄）
River and Lake Waters Start Freezing in the First Pentad of Beginning of Winter (Photo by Yang Jin)

立冬二候：地始冻（杨晋 摄）
The Earth Starts Freezing in the Second Pentad of Beginning of Winter (Photo by Yang Jin)

立冬三候：雉入大水为蜃（杨晋 摄）
Pheasants Enter the Seas and Change into Clams in the Third Pentad of Beginning of Winter (Photo by Yang Jin)

立冬时节，东北农林作物进入越冬期，江淮地区"三秋"已接近尾声，华南地区是"立冬种麦正当时"的最佳时期。

In the solar term of Beginning of Winter, agricultural and forest crops in northeast China enter the wintering period. Autumn harvest, plowing and planting are coming to an end in the reaches of the Yangtze and Huaihe Rivers. In south China, this is the optimal time for sowing wheat.

古时此日，天子有出郊迎冬之礼，并有赐群臣冬衣、矜恤孤寡之制。浙江绍兴，这天要祭祀"酒神"，开酿黄酒。湖南醴陵人要开始熏制有名的"醴陵焙肉"。立冬民间各地还有补冬、吃水饺、冬泳等习俗。

On the day of Beginning of Winter, the Son of Heaven observed the practice of ushering in winter from the outskirts of the capital city, bestowing gift winter clothes to his top officials, and issuing relief to the childless and widowed old people. In Shaoxing, Zhejiang Province, people worshipped "god of wine" and started brewing yellow wine. In Liling, Hunan Province, people started smoking the famed "Liling broiled meat". At Beginning of Winter, people also had the customs of taking in more nutrition for winter, eating dumplings, and winter swimming.

立冬是非常重要的节气，气候变冷，是人们进补的大好时机。中医认为，立冬后阳气潜藏，阴气盛极。万物活动趋向休止，以冬眠状态养精蓄锐，为来年春天生气勃发做准备。

As a very important solar term, Beginning of Winter provides a great opportunity to people to make up for, and supplement, their nutrition at a time when the climate turns cold. According to traditional Chinese medicine, after Beginning of Winter, the *yang* energy starts to conceal itself while the *yin* energy prospers and dominates. Every living thing has a tendency of rest and inactivity. They enter hibernation to preserve their energy, and prepare for blooming back into vitality in the coming spring.

小雪
Light Snow

小雪（韩翠芝 摄）
Light Snow (Photo by Han Cuizhi)

小雪，二十四节气中的第二十个节气，公历每年 11 月 22～23 日，太阳到达黄经 240°时。小雪表示降雪开始的时间和程度，意思是开始降雪，但降雪量还不大，尚未形成积雪。

Light Snow, the twentieth of the Twenty-Four Solar Terms, falls on November 22-23 each year on the Gregorian calendar, when the sun reaches celestial longitude 240°. "Light Snow" refers to the time and degree of snowfall; although snow starts to fall, the amount of snowfall is small, and there is no snow retention on the ground.

小雪三候：一候虹藏不见，二候天气上升地气下降，三候闭塞而成冬。意思是说，此时由于不再有雨，彩虹便不会出现了，同时由于天空中的阳气上升，地中的阴气下降，导致天地不通，阴阳不交，所以万物失去生机，天地闭塞而转入严寒的冬天。

Here are the three pentads of Light Snow: in the first pentad, rainbows are nowhere to be seen; in the second, *yang* energy rises and *yin* energy descends; and in the third, heaven and earth are blocked and winter starts.

二十四节气
The Twenty-Four Solar Terms

小雪一候：虹藏不见（杨晋 摄）
Rainbows are Nowhere to Be Seen in the First Pentad of Light Snow
(Photo by Yang Jin)

小雪二候：天气上升地气下降（杨晋 摄）
Yang Energy Rises and *Yin* Energy Descends in the Second Pentad of Light Snow (Photo by Yang Jin)

小雪三候：闭塞而成冬（杨晋 摄）
Heaven and Earth Are Blocked and Winter Starts in the Third Pentad of Light Snow (Photo by Yang Jin)

小雪节气，中国大部分地区的农业生产已进入冬季田间管理和农田基本建设阶段。小雪日，民间有制作腊肉，南方有吃糍粑，土家族有杀年猪、吃刨汤庆祝丰收等习俗。

In the solar term of Light Snow, agricultural production in most parts of China enters the winter phase of field attendance and basic farm field construction.Customs on the day of Light Snow include smoking pork; eating glutinous rice cakes in south China; butchering pigs for the Spring Festival, and eating fresh pork and drinking fresh pork soup in celebration of the harvest among the ethnic Tu people.

小雪节气前后，天气通常是阴冷晦暗的。从中医角度来讲，此时身体内循环正处于阴盛阳衰的阶段。要多食清火降气的食物，如白萝卜、白菜等当季食物；要注意保暖，减轻压力，放松心情；对于患有心脑血管疾病的人来说，要注意保护心脏。

Around the solar term of Light Snow, the weather is generally somber and gloomy. According to traditional Chinese medicine, the internal circulation of the human body is at a phase of prosperity of the *yin* energy and the decline of *yang* energy. More seasonal foods like white turnips and Chinese cabbages, which have the function of clearing internal heat and reducing pneuma, should be eaten. People should keep warm, reduce pressure, and lighten their moods. For those with cardiovascular and cerebrovascular diseases, their priority lies in the protection of the heart.

大雪
Heavy Snow

大雪(李冬梅 摄)
Heavy Snow (Photo by Li Dongmei)

　　大雪，二十四节气中的第二十一个节气，公历每年 12 月 7 ～ 8 日，太阳到达黄经 255°时。 此时降雪天数和降雪量比小雪节气增多，地面渐有积雪，范围也广，故名大雪。

Heavy Snow, the twenty-first of the Twenty-Four Solar Terms, falls on December 7-8 each year on the Gregorian calendar, when the sun reaches celestial longitude 255°. At this time, the number of snowy days and the amount of snowfall is increased. Snow starts to be accumulated on the ground, and the extent enlarges, thus the name "Heavy Snow".

　　大雪三候：一候鹖鴠不鸣，二候虎始交，三候荔挺出。意思是说此时因天气寒冷，寒号鸟也不再鸣叫了；此时是阴气最盛时期，所谓盛极而衰，阳气已有所萌动，老虎开始有求偶行为；一种名叫"荔挺"的兰草，感到阳气的萌动而抽出新芽。

Here are the three pentads of Heavy Snow. In the first pentad, flying squirrels stop singing; in the second pentad, tigers start courtship; in the third pentad, bonesets sprout. In other words, due to the cold weather, flying squirrels no longer sing; this is the season when the *yin* energy prospers, but since anything extremely prosperous will start to decline, the *yang* energy is already germinating, and thus tigers start their courtship; a boneset named *"liting"* in Chinese feels the germination of the *yang* energy and start to give off new sprouts.

二十四节气
The Twenty-Four Solar Terms

大雪一候：鹖鴠不鸣（杨晋 摄）
Flying Squirrels Stop Singing in the First Pentad of Heavy Snow (Photo by Yang Jin)

大雪二候：虎始交（杨晋 摄）
Tigers Start Courtship in the Second Pentad of Heavy Snow (Photo by Yang Jin)

大雪三候：荔挺出（杨晋 摄）
Boneset Starts Sprouting in the Third Pentad of Heavy Snow (Photo by Yang Jin)

大雪时节的降雪量虽然增长，但是全国大部分地区的降水量却减少了，天气较为干燥，这样的气候为农作物创造了极佳的条件。

While the amount of snowfall increases in this solar term, the precipitation in most parts of China drops. The weather is fairly dry, and this climate creates very favorable conditions for agricultural crops.

大雪时节，江苏南京有腌制咸货、山东有喝红黏粥等习俗。

In this solar term, people have the custom of pickling salty foodstuffs in Nanjing, Jiangsu Province, and people have the custom of eating red glutinous congee in Shandong Province.

大雪节气之后，寒冬来临。冬天的很多季节性疾病都和寒气入侵有关，中医认为，人体的头、胸、脚这三个部位最容易受寒冷侵袭，是保暖的重点。此时要做好内在阳气的保养和封藏工作，尽量减少消耗。养生重点是"补"和"藏"，既要补得进，还要藏得住。

After the solar term of Heavy Snow, the cold winter is here. Many diseases of the season have much to do with the invasion of cold air. According to traditional Chinese medicine, the human head, chest and feet are the three parts most susceptible to the attack of extreme cold, and thus they need more attention and protection. One must preserve, conceal and stock one's own internal *yang* energy, and reduce bodily consumptions. The priorities in health building at this time is "supplement" and "hoarding", and thus one must supplement his *qi* and also hoard his *qi*.

冬 至

Winter Solstice

冬至（高成军 摄）

Winter Solstice (Photo by Gao Chengjun)

冬至，二十四节气中的第二十二个节气，公历每年的12月21～22日，太阳到达黄经270°时。冬至日太阳直射南回归线，北半球昼最短、夜最长，因此"冬至"又叫"日短至"，是二十四节气中最早被测定的节气之一。

Winter Solstice, the twenty-second of the Twenty-Four Solar Terms, falls on December 21-22 on the Gregorian calendar each year, when the sun reaches celestial longitude 270°. On the day of Winter Solstice, the sun shines directly on the Tropic of Capricorn, daylight being the shortest and nighttime longest on the northern hemisphere. Thus, the day is also known as "Shortest Daylight Day", one of the twenty-four solar terms measured and identified the earliest.

冬至三候：一候蚯蚓结，二候麋角解，三候水泉动。意思是说，传说中蚯蚓是阴曲阳伸的生物，此时阳气虽已生长，但阴气仍然十分强盛，土中的蚯蚓仍然蜷缩着身体；麋与鹿同科，却阴阳不同，古人认为麋的角朝后生，所以为阴，而冬至一阳生，麋感阴气渐退而解角；由于阳气初生，所以此时山中的泉水可以流动并且温热。

The three pentads of Winter Solstice are like this. In the first pentad, earthworms curl up; in the second, elks shed their horns; in the third, spring water flows. In other words, according to legends, earthworms coil themselves when the *yin* energy prevails and stretch themselves when the *yang* energy dominates; although at this time the *yang* energy is building up, the *yin* energy is still predominant, and thus earthworms in the earth are still coiled; while elks and deer are of the same family, they have very different *yin-yang* natures; ancient Chinese believed that since the elk horns point backwards, they are of the *yin* energy; at Winter Solstice, the *yang* energy builds up, elks feel the recession of the *yin* energy and thus shed their horns; as the *yang* energy is already building up at its start, the spring water in the mountains flows and is lukewarm.

二十四节气
The Twenty-Four Solar Terms

冬至一候：蚯蚓结（杨晋 摄）
Earthworms Curl up in the First Pentad of Winter Solstice (Photo by Yang Jin)

冬至二候：麋角解（杨晋 摄）
Elks Start to Shed Their Horns in the Second Pentad of Winter Solstice (Photo by Yang Jin)

二十四节气
The Twenty-Four Solar Terms

冬至三候：水泉动（杨晋 摄）
Spring Water Flows in the Third Pentad of Winter Solstice (Photo by Yang Jin)

冬至前后是兴修水利、大搞农田基本建设、积肥造肥的大好时机。

It is the best time to build water conservancy projects, be engaged in farmfield preparations, and store compost and fertilize manure on a large scale around Winter Solstice.

冬至是中国古代重要节日，天子要率领文武百官举行祭天仪式，民间则要祭祖贺冬、举办冬至亚岁宴等。如今民间还有贴消寒图，吃饺子、冬至圆、赤豆粥、黍米糕，喝冬酿酒等习俗。浙江三门地区会举办三门祭冬活动，迎接新年、庆贺丰收、感恩尽孝。

Winter Solstice is a significant festival in ancient China, on which the Son of Heaven would head the civilian officials and military officers to hold ceremonies to make sacrificial offerings to Heaven. Ordinary people would make offerings to their ancestors and celebrate winter, and hold *yasui* ("secondary spring festival") banquets. Nowadays, there are still folk customs of hanging cold-repelling pictures, eating dumplings, Winter Solstice rice dumplings, red beans congee, millet cakes, and drinking winter-brewed wine. In the Sanmen area of Zhejiang Province, people would hold Winter Worshipping activities, to usher in the New Year, celebrate the bumper harvest, show appreciation, and be filially pious.

冬至一阳生，在这一天，盛到极点的阴气开始衰退，从而会有一点阳气萌生，所以这是阴阳转换的时刻，历代养生家都很重视在这个节气上的养生。此时科学养生有助于保证旺盛的精力，达到延年益寿的目的。

On the day of Winter Solstice, one tinge of *yang* energy has emerged. It is on this day that the *yin* energy at its extreme dominance starts to recede, and thus one small dose of *yang* energy will germinate. Therefore, this is a moment of *yin-yang* shift. Health building experts of various dynasties gave much emphasis to this solar term. At this time, if one preserves his health scientifically, he will go a long way in keeping up his vibrant vigor and lengthening his life span.

小 寒
Lesser Cold

小寒（尹涛 摄）
Lesser Cold (Photo by Yin Tao)

二十四节气
The Twenty-Four Solar Terms

小寒，二十四节气中的第二十三个节气，公历每年1月5～7日，太阳到达黄经285°时。标志着中国大部地区开始进入一年中最寒冷的时段。

Lesser Cold, the twenty-third of the Twenty-Four Solar Terms, falls on January 5-7 on the Gregorian calendar each year, when the sun reaches celestial longitude 285°. This is the indication that most parts of China have entered the coldest period of the year.

小寒三候：一候雁北乡，二候鹊始巢，三候雉雊。意思是说古人认为候鸟中大雁是顺阴阳而迁移的，此时阳气已动，所以大雁开始向北迁移；喜鹊因感觉到阳气而开始筑巢；雉在接近四九时会因感阳气的生长而鸣叫。

The three pentads of Lesser Cold are like this. In the first pentad, migrant swan geese return to the north; in the second pentad, magpies start building their nests; in the third pentad, pheasants start chuckling. In other words, ancient Chinese believed that among the migrant birds, swan geese carry out their migration in line with the *yin-yang* shift; as the *yang* energy starts growing at this time, swan geese start to move to the north; magpies feel the *yang* energy and start building nests for themselves; pheasants also feel the growth of the *yang* energy and chuckle near the time of the fourth nine-day cold period after Winter Solstice.

二十四节气
The Twenty-Four Solar Terms

小寒一候：雁北乡（杨晋 摄）
Swan Geese Return to Their Northern Homes in the First Pentad of Lesser Cold (Photo by Yang Jin)

二十四节气
The Twenty-Four Solar Terms

小寒二候：鹊始巢（杨晋 摄）
Magpies Start Building Their Nests in the Second Pentad of Lesser Cold (Photo by Yang Jin)

小寒三候：雉雊（杨晋 摄）
Pheasants Start Chuckling in the Third Pentad of Lesser Cold (Photo by Yang Jin)

小寒时节温度很低，对农作物的危害较大，要做好农作物的防寒防冻工作。旧时天津地区有小寒日吃黄芽菜的习俗，南京地区是吃菜饭，广州地区则吃糯米饭，民间还有在这一天补膏方等习俗。

During the Lesser Cold period, the temperature is very low, with great damaging effect on agricultural crops. It is thus imperative to carry out the anti-cold and anti-freezing measures for the crops. In the old times in Tianjin, there was the custom of eating Chinese cabbage on the day of Lesser Cold; in Nanjing, people had vegetable rice meals; in Guangzhou, people had glutinous rice meals; some people might also make a point of taking herbal tonic paste on this day.

寒为冬季的主气，小寒又是一年中最冷的时节。寒为阴邪，易伤人体阳气，寒主收引凝滞。所以在小寒节气里，患心脏病和高血压病的人往往会病情加重，患脑卒中者增多，尤其是老年人一定要做好保暖。

Cold is the dominant ambience in winter, and Lesser Cold is the coldest season in the year. As a *yin* evil, cold might hurt the *yang* energy in the human body easily, by inducing spasms and stagnation. Therefore, in this solar term, those with heart diseases and high blood pressure may experience an exasperation of their condition, with an increased number of cerebral apoplexy cases; thus, elderly people, in particular, must keep warm.

大 寒
Greater Cold

大寒（杨晋 摄）
Greater Cold (Photo by Yang Jin)

　　大寒，二十四节气中的最后一个节气，公历每年 1 月 20 日前后，太阳到达黄经 300° 时。大寒是相对小寒而言，是一年中最寒冷的时期。

Greater Cold, the last one of the Twenty-Four Solar Terms, falls around January 20 on the Gregorian calendar each year, when the sun reaches celestial longitude 300°. Greater Cold is so named in comparison with Lesser Cold, being the coldest time of the year.

　　大寒三候：一候鸡乳，二候征鸟厉疾，三候水泽腹坚。意思是说，此时母鸡开始孵养小鸡了；而鹰隼之类的征鸟，却正处于捕食能力极强的状态中，盘旋于空中到处寻找食物，以补充身体的能量抵御严寒；水中的冰一直冻到水中央，且最结实、最厚。

Here are the three pentads of Greater Cold. In the first pentad, hens start hatching eggs; in the second pentad, birds that fly long distances exert themselves; in the third pentad, rivers and lakes are frozen to the very heart of the water body. In other words, at this time, hens start to hatch their eggs for chicks; birds that fly long distances, such as hawks and falcons, in a state of active and effective prey, are circling high up in the sky and looking out for targets, in order to compensate the energy in the body against the extreme cold; ice in rivers and lakes is frozen to the depth of the water body, with the greatest solidness and thickness in the year.

大寒一候：鸡乳（杨晋 摄）
Hens Start to Hatch Eggs in the First Pentad of Greater Cold (Photo by Yang Jin)

大寒二候：征鸟厉疾（杨晋 摄）
Long-Flying Birds Exert Themselves in the Second Pentad of Greater Cold (Photo by Yang Jin)

二十四节气
The Twenty-Four Solar Terms

大寒三候：水泽腹坚（杨晋 摄）
Water Bodies on Land Froze to the Depth in the Third Pentad of Greater Cold (Photo by Yang Jin)

北方地区老百姓多忙于积肥堆肥，为开春做准备，或者加强牲畜的防寒防冻。南方地区则仍加强小麦及其他作物的田间管理。大寒时节也是岭南当地集中消灭田鼠的重要时机。

People in north China are generally busy in collecting manure and stockpiling compost to get ready for the spring season, or they engage themselves in enhancing the anti-cold and anti-freezing measures for livestock. In south China, people step up attendance of wheat and other crops in the field. During the Greater Cold season, it is the best time to terminate field mice in Lingnan areas.

广东佛山民间有大寒节用瓦锅蒸煮糯米饭的习俗，安徽安庆则有大寒炸春卷的习俗。此外在一些地方，大寒日还有尾牙祭、买芝麻秸等习俗。

In Foshan, Guangdong Province, people have the custom of cooking glutinous rice meals in large pottery cookers on the day of Greater Cold. In Anqing, Anhui Province, there is the practice of frying spring rolls. In other places on the day of Greater Cold, there are the customs of end-of-the-year God of Earth worship and celebrations, and of purchasing sesame straws.

养生方面，冬三月是生机潜伏、万物蛰藏的时令，此时人体的阴阳消长代谢也处于相当缓慢的时候，所以应该早睡晚起，不要轻易扰动阳气，凡事不要过度操劳，要使神志深藏于内，避免急躁发怒。

In terms of health building, the three months of winter are a season of vigor and vitality lying latent, and living things staying inconspicuous and inactive. The human body is in a state of very slow rise and fall, and very gradual metabolism of the *yin* and *yang* energy inside. Therefore, one should go to bed early and get up late. One had better not disturb the *yang* energy without a good reason. One is advised against too much exertion, so that the mind and physical energy should be well concealed and one should avoid losing one's temper or flying into a rage.

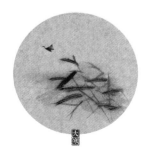

结 语
Concluding Remarks

  二十四节气作为中华优秀传统文化的组成部分，引领着中华传统农业从奠基走向辉煌，为中华民族的生存和繁衍发挥着重要作用，其所蕴含的"天人合一"哲学理念，体现了人与人和睦相处、人与自然和谐共生的"和合"文化精神，不仅培育了中国人尊重自然规律和生命节律的世界观，也塑造了天道均平、以和为贵的社会生活理想。二十四节气对促进中国经济社会全面、协调、可持续的和谐发展发挥着积极作用，对于顺应自然规律变化，指导农业遵循规律，趋利避害，应时而作，顺时而为也起到积极有效的调适作用。因此，保护和传承二十四节气不仅具有深远的历史意义，也具有重要的现实意义。

二十四节气
The Twenty-Four Solar Terms

As part of the best Chinese traditional culture, the "Twenty-Four Solar Terms" ushered traditional Chinese agriculture from its roots to splendor, playing a significant role in the survival and reproduction of the Chinese race. The philosophical concept of "harmony between man and heaven" reflects the "peaceful and harmonious" cultural spirit that people should live in peace with each other, and that man and nature should co-exist in harmony. This not only cultivates in the Chinese a world view that natural rules and principles as well as life rhythms should be respected, but also shapes an ideal for social life that heaven above is fair and impartial to all and that peacefulness is prized. The "Twenty-Four Solar Terms" have been playing an active role in promoting the comprehensive, coordinated, sustainable and harmonious development of the Chinese economy and society. It has also been playing an active and effective role of accommodation in conforming to natural rules and principles, guiding agriculture to follow the nature's rules, going for benefits and avoiding harms, farming according to the natural seasonal requirements, and abiding by the best time of doing things. Therefore, the protection and inheritance of the "Twenty-Four Solar Terms" has both a far-reaching historical significance, and an important realistic implication.

"春雨惊春清谷天，夏满芒夏暑相连，秋处露秋寒霜降，冬雪雪冬小大寒。"愿这首千百年来世代传唱的"二十四节气歌"，永远与时光相随，与生命相伴。

Beginning of Spring, Rain Water, Insects Awaken, Spring Equinox, Freshgreen, Grain Rain, being the spring season.
Beginning of Summer, Lesser Fullness, Grain in Ear, Summer Solstice, and Lesser and Greater Heat, being the summer season.
Beginning of Autumn, End of Heat, White Dew, Autumn Equinox, Cold Dew, and First Frost, these are the solar terms of autumn.
Beginning of Winter, Light and Heavy Snow, Winter Solstice, Lesser and Greater Cold are thus the solar terms of winter.
This "Song of the Twenty-Four Solar Terms" that has been passed down from generation to generation for thousands of years, shall exist as long as time, and serve to be company to life forever.

附录
Appendix

保护和传承"人类非物质文化遗产"二十四节气倡议书

An Initiative on the Protection and Preservation of the "Twenty-Four Solar Terms — An Intangible Cultural Heritage of Humanity"

2016年11月30日,联合国教科文组织保护非物质文化遗产政府间委员会第十一次常会正式将"二十四节气"——中国人通过观察太阳周年运动而形成的时间知识体系及其实践列入联合国教科文组织人类非物质文化遗产代表作名录。

二十四节气
The Twenty-Four Solar Terms

On November 30, 2016, the 11th session of the UNESCO's Intergovernmental Committee for the Safeguarding of Intangible Cultural Heritages proclaimed "China—The Twenty-Four Solar Terms, knowledge of time and practices developed in China through observation of the sun's annual motion" a Masterpiece of the Intangible Cultural Heritage of Humanity.

　　二十四节气是中国人通过观察太阳周年运动，认知一年中时令、气候、物候等方面变化规律所形成的知识体系和社会实践，是农耕社会生产生活的时季指南，是民族生存发展的文化时间。作为中国人特有的时间认知体系，二十四节气深刻影响着人们的生产方式、生活方式、思维方式和行为准则，是中华民族文化认同的重要载体，是全人类共同的文化财富！

The twenty-four solar terms are a knowledge system and social practice of the Chinese people that have evolved from their observations of the solar movements of the year, and from their perception and understanding of the rules of the changes in the seasons, climate and phenology throughout the year. They are the seasonal guides of the production and life in the agricultural society, and also the cultural time of the existence and development of the nation. As the system of time cognition unique to the Chinese, the twenty-four solar terms have profoundly influenced people's ways of production, life and thinking as well as their code of conduct, serving as a major carrier of the identity of the Chinese nation and culture, and the collective cultural assets of the whole human race.

　　二十四节气作为人类非物质文化遗产，是中国五千年灿烂文明及非凡创造力的集中体现与智慧结晶；是中国农耕历史发展和人类社会进步的永恒记忆；更是我们传承历史、继往开来、发扬光大的文化渊

源和动力。我们要像保护好我们的眼睛一样保护好这一项人类非物质文化遗产,动员全社会力量,采取更加有组织、有计划、有针对性的措施,通过我们的科学保护和有效传承,使二十四节气知识体系和系列文化得到更好地弘扬,使这一中国优秀传统文化得到更多中国人关注,成为中国人日常生活的重要组成部分,并与人类共享。

As intangible cultural heritage of humanity, the twenty-four solar terms are a concentrated reflection of the splendid five-thousand-year Chinese civilization and the country's remarkable creativity, and they are also the precipitation of wisdom in this process. They are now a permanent memory of the historical development of the Chinese agriculture and farming and of human social progress. They are also a cultural source and a driving force in our inheritance of history, in our march into the future and in our noble endeavors for greater brilliance and magnificence. We should protect the intangible cultural heritage of humanity like the pupils of our eyes, mobilize forces of the whole society, adopt more organized, more planned and more targeted measures, and better promote the knowledge system and cultural series of the twenty-four solar terms through our scientific protection and efficient preservation. Through all these efforts, we hope to enable more Chinese to pay attention to this prime Chinese traditional culture, render it into an important component of the daily life of Chinese people, and share it with the rest of mankind.

为传承五千年灿烂文明,保护和继承先民的智慧结晶,推进"二十四节气"区域间文化和信息的协作与共享,中国农业博物馆、中国民俗学会与浙江省衢州市柯城区等十二个相关群体、社区,共同向社会各界和广大民众发出如下倡议:

With the aim of preserving the splendid civilization of five thousand years, protecting and inheriting our ancestors' fruition of wisdom, and enhancing the cultural and informational coordination and sharing between the "twenty-four-solar-term" regions, twelve institutions and communities including the China Agricultural Museum, China Folklore Society, and Kecheng District of Quzhou Municipality, Zhejiang Province, launch the following initiative to the various spheres of society and the general public.

1. 要进一步了解和学习二十四节气传统文化相关知识，从中体悟古代中国人探索自然奥秘、实现人与自然和谐统一的求知精神和生活智慧。

1. Further understand and study the traditional cultural knowledge of the twenty-four solar terms, and reflect on the curiosity for knowledge and the life wisdom of the ancient Chinese in their exploration of the mysteries of the nature and in their pursuit of the harmony and unity between man and nature.

2. 要亲历和参与家乡二十四节气民间风俗，记录关于节气的民谣、农谚、传说、故事等口头文学样式，记录关于节气的生活礼仪、饮食文化样式，记录当地关于节气的时令养生、生活宜忌等习惯，了解并记录与节气有关的器物等，使这些民俗文化样式以文字、图片、视频等形式得以留存。

2. Experience and participate in the folk customs of the twenty-four solar terms held in your hometowns; record solar-term related oral literature of ballads, agricultural proverbs, legends and stories; note down the daily life rituals and cuisine cultural varieties having to do with the solar terms; note down local practices and customs regarding seasonal health building and daily do's and don'ts relevant to the solar terms; learn about, and keep note of, utensils and artifacts related to the solar terms, so that these folk cultural varieties shall be protected and preserved in the form of writings, graphics and videos.

3. 要向后代、晚辈讲解二十四节气相关知识、风俗文化等，促进代际传承。

3. Pass on the relevant knowledge, folk customs and culture in relation to the twenty-four solar terms to your children and children's children and other younger relatives and acquaintances, in an effort to promote its inheritance through the generations.

4. 要向境外友人推介以二十四节气为代表的中国智慧。

4. Promote and advocate Chinese wisdom represented by the twenty-four solar terms, to friendly personnel overseas.

让我们共同携手，在继承的基础上，将先民文化遗产发扬光大，让中华文明的结晶在当代和将来迸发出更加灿烂、更加耀眼的光华！为建设社会主义文化强国，增强国家文化软实力，为实现中华民族伟大复兴的中国梦而共同努力！

Let us join hands, and based on preservation, promote and advocate the cultural legacy of our ancestors! Let the fruit of Chinese civilization shine with greater splendor and brilliance at present and in future! Let us make our concerted efforts in building a socialist cultural power, enhancing the national cultural soft power, and achieving the Chinese dream of the great revitalization of the Chinese nation!

图书在版编目（CIP）数据

二十四节气／中国农业博物馆组编．—北京：中国农业出版社，2019.5

（"人类非物质文化遗产代表作——二十四节气"科普丛书）

ISBN 978-7-109-25403-9

Ⅰ．①二… Ⅱ．①中… Ⅲ．①二十四节气－风俗习惯－中国－通俗读物 Ⅳ．① P462-49 ② K892.18-49

中国版本图书馆 CIP 数据核字（2019）第 061792 号

中国农业出版社出版
（北京市朝阳区麦子店街18号楼）
（邮政编码100125）
责任编辑　张德君　李　晶　司雪飞
文字编辑　杨　春

北京中科印刷有限公司印刷　新华书店北京发行所发行
2019年6月第1版　2019年6月北京第1次印刷

开本：787mm×1092mm　1/16　印张：9.5
字数：190千字
定价：68.00元
（凡本版图书出现印刷、装订错误，请向出版社发行部调换）